Lecture Notes in Computer Scie

Edited by G. Goos, J. Hartmanis and J. van L

Springer
Berlin
Heidelberg
New York
Barcelona
Hong Kong
London
Milan
Paris
Singapore
Tokyo

Stephen W. Liddle   Heinrich C. Mayr
Bernhard Thalheim (Eds.)

Conceptual Modeling for E-Business and the Web

ER 2000 Workshops on
Conceptual Modeling Approaches for E-Business and
The World Wide Web and Conceptual Modeling
Salt Lake City, Utah, USA, October 9-12, 2000
Proceedings

Springer

Series Editors

Gerhard Goos, Karlsruhe University, Germany
Juris Hartmanis, Cornell University, NY, USA
Jan van Leeuwen, Utrecht University, The Netherlands

Volume Editors

Stephen W. Liddle
Brigham Young University
Marriott School, School of Accountancy and Information Systems
585 TNRB, P.O. Box 23087, 84602-3087 Provo, Utah, USA
E-mail: liddle@byu.edu

Heinrich C. Mayr
University of Klagenfurt
Insitute for Business Informatics and Application Systems
Universitätsstr. 65-67. 9020 Klagenfurt, Austria
E-mail: heinrich@ifit.uni-klu.ac.at

Bernhard Thalheim
Brandenburg University of Technology at Cottbus
Computer Science Institute
Postfach 101344, 03013 Cottbus, Germany
E-mail: thalheim@informatik.tu-cottbus.de

Cataloging-in-Publication Data applied for

Die Deutsche Bibliothek - CIP-Einheitsaufnahme

Conceptual modeling for E-business and the web : proceedings / ER 2000 ;
Workshops on Conceptual Modeling Approaches for E-Business and the
World Wide Web and Conceptual Modeling, Salt Lake City, Utah, USA,
October 9 - 12, 2000. Stephen W. Liddle ... (ed.). - Berlin ;
Heidelberg ; New York ; Barcelona ; Hong Kong ; London ; Milan ; Paris ;
Singapore ; Tokyo : Springer, 2000
    (Lecture notes in computer science ; Vol. 1921)
    ISBN 3-540-41073-2

CR Subject Classification (1998): H.2, H.3-5, C.2.4-5, J.1

ISSN 0302-9743
ISBN 3-540-41073-2 Springer-Verlag Berlin Heidelberg New York

Springer-Verlag Berlin Heidelberg New York
a member of BertelsmannSpringer Science+Business Media GmbH
© Springer-Verlag Berlin Heidelberg 2000
Printed in Germany

Typesetting: Camera-ready by author, data conversion by PTP-Berlin, Stefan Sossna
Printed on acid-free paper   SPIN: 10722824   06/3142   5 4 3 2 1 0

Preface

The objective of the workshops associated with the ER2000 19th International Conference on Conceptual Modeling was to give participants the opportunity to present and discuss emerging, hot topics, thus adding new perspectives to conceptual modeling. This attracts communities which have begun to or which have already recognized the importance of conceptual modeling for solving their problems. To meet this objective, we selected the following two topics:

- Conceptual Modeling Approaches for E-Business (eCOMO2000) aimed at studying the application of conceptual modeling techniques specifically to e-business.
- The World Wide Web and Conceptual Modeling (WCM2000) which analyzes how conceptual modeling can help address the challenges of Web development, management, and use.

eCOMO2000 is the first international workshop on Conceptual Modeling Approaches for E-Business. It was intended to work out and to discuss the actual state of research on conceptual modeling aspects and methods within the realm of the network economy, which is driven by both traditionally organized enterprises and dynamic networks. Following the philosophy of the ER workshops, the selection of eCOMO contributions was done very carefully and restrictively (six accepted papers out of thirteen submissions) in order to guarantee an excellent workshop program. We are deeply indebted to the authors and to the members of the program committee, whose work resulted in this outstanding program. The contributions are organized into two sessions:

- *Modeling Approaches.* Clearly, e-business results in new challenges for information system developers in supplying Web-based services which will go far beyond existing notions like B2B (business-to-business) and B2C (business-to-consumer). This comes with new aspects and must be covered by modeling methods and techniques. The papers in this session discuss such aspects, e.g. the necessary ontology-based harmonization of Web content standards, the modeling of process-centered, highly complex systems, and the use of hypermedia modeling techniques for the conceptual design of electronic product catalogs.
- *Modeling E-Business Processes and Workflow Markets.* In the context of modeling, process modeling is an important issue. Processes are the "meat" of e-business. Common visual-oriented process modeling concepts may lead to "spaghetti process" models. Within this session an alternative approach leading to nested logic units according to the concept of linear programming is presented. Another contribution critically analyzes the widespread opinion that e-business models are similar to process models. It argues for a different position that views e-business as the exchange of values between actors. Last but not least, modeling aspects of electronic workflow markets are discussed.

WCM2000 is the second international workshop on the World Wide Web and Conceptual Modeling, following a very successful first workshop at ER'99. Interest this year was again strong. We received twenty-three submissions and accepted nine. We express our appreciation to the authors and program committee whose hard work made this excellent program possible. The papers are organized into three sessions:

- *Web Application Modeling I and II.* The current state of the Web clearly demonstrates the need for more organized approaches to Web application modeling. The papers in these sessions propose tools and techniques to improve the state of the art in Web development. Papers in the first session specifically address methods for developing Web applications that are personalized and support multiple output devices using *WebML* and *W3I3*; methods to improve the maintenance and evolution of Web applications using *OOHDM*; and new issues that arise in a Web context because of its integration of hypermedia and database aspects. Papers in the second session propose *webspaces* for modeling Web data; *user interaction diagrams* for modeling interactions and navigation in Web applications; and *WISDOM*, a framework for developing Web-based information systems.
- *Managing and Querying Web Data and Metadata.* In a Web context, metadata is particularly helpful because it can enable the automated interchange and processing of information. The first paper in this session explores the use of the Resource Description Framework (RDF) standard, and proposes a method for modeling and querying RDF schemas. The second paper tackles the problem from a different perspective, recognizing that much existing unstructured or semistructured data must be managed, and so proposes an example-based environment for wrapper generation. The final paper explores the issue of improving Web queries by flexibly categorizing retrieved documents.

In order to provide the maximum benefit possible, the ER2000 workshops were open to all ER2000 attendees, and workshop sessions were held in parallel with other ER2000 sessions.

We acknowledge the hard work of the many individuals who made these workshops a great success. We appreciate the diligent service of the organizing committees and program committees. We especially acknowledge and thank Christian Kop, who managed many of the administrative details for eCOMO2000.

October 2000

Bernhard Thalheim
Heinrich C. Mayr
Stephen W. Liddle

ER2000 Workshops Organization

General

ER2000 Workshops Chair: Bernhard Thalheim
Brandenburg University of Technology at Cottbus, Germany

eCOMO2000

Workshop Chair: Heinrich C. Mayr
University of Klagenfurt, Austria
Organization: Chrisian Kop
University of Klagenfurt, Austria

WCM2000

Workshop Chair: Stephen W. Liddle
Brigham Young University, USA
Steering Committee Chair: Peter P. Chen
Louisiana State University, USA
Steering Committee: David W. Embley
Brigham Young University, USA

eCOMO2000 Program Committee

Anthony Bloesch	Microsoft Corporation, USA
Hans Ulrich Buhl	University of Augsburg, Germany
Nicola Guarino	National Research Council, Italy
József Györkös	University of Maribor, Slovenia
Bill Karakostas	University of Science and Technology in Manchester, UK
Roland Kaschek	UBS Zurich, Switzerland
Jacques Kouloumdjian	INSA-Lyon Scientific and Technical University, France
August Wilhelm Scheer	University of Saarbrücken, Germany
Bernhard Thalheim	Brandenburg University of Technology at Cottbus, Germany
Roland Wagner	University of Linz, Austria
Christopher Welty	Vassar College/IBM, USA
Carson Woo	University of British Columbia, Canada
Tom Worthington	Australian National University, Australia

WCM2000 Program Committee

Xavier Alamán	Independent University of Madrid, Spain
Terje Brasethvik	Norwegian University of Science and Technology, Norway
Lois Delcambre	Oregon Graduate Institute, USA
Franca Garzotto	Milan Polytechnic, Italy
Bertram Ludaescher	University of California at San Diego, USA
Wolfgang May	Freiburg University, Germany
Paolo Paolini	Milan Polytechnic, Italy
Óscar Pastor	Polytechnic University of Valencia, Spain
Gustavo Rossi	National University of La Plata, Argentina
Daniel Schwabe	Pontifical Catholic University of Rio de Janeiro, Brazil
Altigran Soares da Silva	Federal University of Minas Gerais, Brazil
Il-Yeol Song	Drexel University, USA

External Referees

Irna Marilia R. Evangelista Filha	Jon Atle Gulla
Paulo B. Golgher	Amarnath Gupta
Jaime Gómez	

Table of Contents

Web Application Modeling II

Managing and Querying Web Data and Metadata

Towards Ontology-Based Harmonization of Web Content Standards

Nicola Guarino, Christopher Welty[†], and Christopher Partridge

LADSEB/CNR
Padova, Italy
{guarino,welty}@ladseb.pd.cnr.it
http://www.ladseb.pd.cnr.it/infor/ontology/ontology.html
† on sabbatical from Vassar College, Poughkeepsie, NY

Abstract. The popularity and press surrounding the release of XML has cre-ated widespread interest in standards within particular communities that fo-cus on representing content. The dream is that these standards will enable consumers and B2B systems to more accurately search information on the Web within these communities. We believe the expansiveness and diversity of the Web creates a need for a small set of standard semantic primitives that have the same meaning and interpretation across communities. Such a stand-ard set of primitives should take into account existing efforts in ontology, and in e-commerce content standards. We are investigating existing content standards proposals for the Web, and present here a large, but by no means complete, list of these standards efforts classified by their ontological sophis-tication and their intent. We then propose some very preliminary notions of how these standards could be harmonized to produce a set of standard seman-tic primitives for describing content.

1 Introduction

The popularity and press surrounding the release of XML has created widespread inter-est in standard tagsets within particular communities that focus on representing *content*. The goal of these standards is to enable consumers and B2B systems to more accurately search information on the Web *within these communities*. Each one of these standards efforts is being carried out in almost complete isolation from the others.

Practitioners and researchers in information system and software engineering know full well that no system will function in isolation for very long. Eventually, it seems clear that these communities will require some interoperability. For example, the Auto-mobile Network Exchange (ANX) and the Open Financial Exchange (OFX) may seem at the current time to be independent and isolated communities (indeed, they are each developing their own web standards, see below). It is not hard to imagine, however, that companies in the automobile industry will want to interact with companies in the financial industry. Certainly they do outside of the world of e-commerce, since many people get financing when they buy a car, and many car dealers get financing when they fill their lots.

Unfortunately, even those communities that do recognize the eventual need for interoperability seem to believe that since their content standards are expressed in XML, they will be able to interoperate. We believe that there is a need for a set of stan-dard semantic primitives that have the same meaning and interpretation across commu-

S.W. Liddle, H.C. Mayr, B. Thalheim (Eds.): ER 2000 Workshop, LNCS 1921, pp. 1–6, 2000.
© Springer-Verlag Berlin Heidelberg 2000

nities, and can therefore serve as a starting point in any integration or interoperation effort.

Proposing a set of standard semantic primitives may sound very much like proposing a standard top-level ontology. While there are certainly similarities between the two, the primary difference is that we are working towards *a very small and simple set of primitives* that takes into account existing efforts in ontology, and in e-commerce content standards.

This work is in the preliminary stages, at the moment we are collecting and cataloging content standardization efforts, and attempting to classify them according to a simple scheme we have devised and are evolving. In this position paper we present our classification scheme, followed by an incomplete list of efforts and how they fit into our scheme. We conclude with some brief comments on how we believe this work will impact our ultimate goal to develop a set of standard semantic primitives for the Web.

2   Classifying Standardization Efforts

We have analysed a number of content standardization efforts and identified what are, for our purposes, relevant properties we wish to associate with each one. The following sections describe our classification scheme and explain the entries found in the appendix.

2.1   Main Classifications

All standards efforts are classified along two principle dimensions, the *semantic depth*, and the *domain type.*

Semantic Depth. There are five levels of semantic depth in our ontology, in order of increasing depth:

  0. *Repositories* (Biztalk, EDI, ...)

  1. Interchange formats and *protocols* (XML/EDI, eCo...)

  2. Common data *dictionaries* (ISO/BSR, CALS/UDEF, Dublin Core...)

  3. *Thesauri* (WordNet, EuroWordNet, UN/SPSC...)

  4. Reference *models* (Meta Data Coalition, Workflow Management Coalition, Indecs...)

  5. Axiomatic *ontologies* (CYC, DARPA/CPR, NIST/PSL ontology...)

Domain Type. There are five specific types of domains that have arisen in our analysis:

  1. *General* content standards and ontologies (WordNet, CYC, ISO/BSR, CALS/UDEF, ...)

  2. *Process* standards (NIST/PSL ontology, DARPA/CPR, ...)

  3. *Product* standards (ISO/STEP, UN/SPSC, RosettaNet...)

  4. *Information* media standards (Dublin Core, INDECS, CIDOC...)

  5. Conceptual modeling and *representation* standards (UML meta-model, EPISTLE...)

2.2   Other Attributes

In addition to the primary classifications, and obvious attributes like *acronym*, *name*, and *URL*, we have also recorded certain other information to assist in understanding the standardization efforts.

Meta-Content. Some standardization efforts specifically address meta-content issues, as opposed to simply describing a domain. This is basically a boolean attribute, however some efforts such as the Dublin Core deal with meta-content in such a simplistic fashion that we chose to differentiate it from others with a "some" value.

A few efforts, such as the Basic Semantic Repository, were so vague on the issue we created another value indicating we tried to figure it out but couldn't (?).

Business Content. Most efforts contained business knowledge encapsulated in the standard, while others (normally the efforts classified as having general domain specificity) attempted to be independent. This, like meta-content, is basically a boolean attribute with two possible variations.

Focus. Names the domain of interest, if any, motivating the standardization effort.

3   Content Standardization Efforts

The appendix contains a table listing content standardization efforts we have found that focus on e-business. We make no claims of completeness nor accuracy, the list is being compiled as a resource for ourselves and others, and will be made public on our web page shortly. Blank cells in the tables indicate that we have not yet determined a value for that attribute. A "?" indicates we have tried to determine a value but were unable to, usually because the web pages or other information we found was incomplete.

The table shows content standardization efforts that we have classified according to the two main criteria (domain specificity and semantic depth) discussed in the previous section.

4   Towards a Standard Semantics

Our intent is to study these standardization efforts, and develop a small set of semantic primitives that seem most likely to have meanings that can be agreed upon and standardized. These primitives will serve as primitive *semantic* tools, the way XML serves as a primitive *syntactic* tool, and will be described and standardized using basic notions of formal ontology [2] such as those described in [3].

Examples of likely candidates for this standardization are high-level notions such as *Part* [1], and *Subject* [4], and lower level notions such as *Person* and *Document*.

Our goal will not be to create a monolithic top-level ontology that specifies one way of viewing the world, but to create a minimal set of primitives that have a crisp and agreed-upon meaning.

5   Conclusion

We have presented a simple ontology for classifying content standardization efforts that organizes them along two primary dimensions: *semantic depth* and *domain specificity*, and then presented a significant number of efforts in the framework of our ontology. We hope this will prove to be an important resource for content standardization efforts.

We are using this resource as input to a proposal for a minimal set of standard semantic primitives that can augment the standard syntax primitives offered by the existing XML standard.

References

1.   Artale, A., Franconi, E., Guarino, N., and Pazzi, L. 1996. Part-Whole Relations in Object-Centered Systems: an Overview. *Data and Knowledge Engineering*, **20**(3): 347-383. Elsevier.
2.   Guarino, N. 1998. Formal Ontology in Information Systems. In N. Guarino (ed.) *Formal Ontology in Information Systems*. Proceedings of FOIS'98, Trento, Italy, 6-8 June 1998. IOS Press, Amsterdam: 3-15.
3.   Guarino, N., and Welty, C. 2000. Identity, Unity, and Individuality: Towards a Formal Toolkit for Ontological Analysis. In, *Proceedings of ECAI-2000: The European Conference on Artificial Intelligence*. Available from http://www.ladseb.pd.cnr.it/infor/ontology/Papers/OntologyPapers.html.
4.   Welty, C. and Jenkins, J. 1999. An ontology for subject. *Data and Knowledge Engineering*. **31**(2):155-182. Elsevier.

Table of Content Standardization Efforts

Acronym	Full Name	Semantic Depth	Domain Type	Focus	Meta Cont.	Bus. Cont.	Web address
AAT	Art and Architecture Thesaurus	Thesaurus	Product	Art and architecture	No	Yes	http://shiva.pub.getty.edu/aat_browser/
ANX	Automobile Network eXchange	Thesaurus	Product	Automotive	No	Yes	http://www.anxo.com/
BizTalk		Protocol	Process	Business documents handling		Yes	http://www.biztalk.org/biztalk/default.asp
CDIF	Common Data Interchange Framework	Model	General	Electronics	Yes	—	http://www.eigroup.org/cdif/index.html
CIDOC	International Committee for Documentation	Model	Product	Museums	?	Yes	http://www.cidoc.icom.org/guide
CIMI profile	Z39.50 Profile for Cultural Heritage Information	Thesaurus	Product	Museums	No	Yes	http://www.cimi.org/
CWM	Common Warehouse Model	Model	Information	Data Warehousing	No?	No	http://www.cwmforum.org/
cXML	Commerce XML	Protocol	Process	eCommerce	No	No	http://www.cxml.org/home/
DCMI	Dublin Core Metadata Initiative	Model	Information	Libraries/Web Catalogs	No	Yes	http://purl.org/dc/
EDIFACT	Electronic Data Interchange for Administration, Commerce and Trade	Protocol	Product	eCommerce	No	Yes	http://www.edifact-wg.org/
EPISTLE	European Process Industries STEP Technical Liaison Executive	Model	Representation	Process industry	Yes	Yes	http://www.step-com.ncl.ac.uk/epistle/epistle.htm
FpML	Financial products Markup Language	Protocol	Representation	Finance	No	Yes	http://www.fpml.com/ http://www.fpml.org/
HL7	Health Level Seven	Model	Product	Healthcare	Yes	Yes	http://www.hl7.org/
ICE	Information & Content Exchange	Protocol	Process	Exchange of online assets	?	Yes	http://www.oasis-open.org/cover/ice.html
IDEF5	Integration Definition for Function Modeling Ontology Description Capture	Model	Representation	Methodologies	Yes	No	http://www.idef.com/
INDECS	interoperability of data in e-commerce systems	Model	Information	Intellectual Property	No	Yes	http://www.indecs.org
ISITC	International Securities Association for Institutional Trade Communication	Model	Representation	Security Industry Transactions	No	Yes	http://www.isitc.org/
ISO 2789	Documentation--Guidelines for the establishment and development of monolingual thesauri	Dictionary	General	Thesauri	Yes	No	
ISO IEC 11179	Metadata Registry Coalition	Model	General	Modelling	Yes	No	http://www.sdct.itl.nist.gov/~ftp/l8/other/coalition/Coalition.htm
ISO1087	Terminology – Vocabulary	Dictionary	General	Terminology	Yes	No	
ISO5964	Documentation and establishment of multilingual thesauri	Dictionary	General	Thesauri	Yes	No	
MDCOIM	MDC Open Information Model	Model	General		Yes	Yes	http://www.mdcinfo.com/
MOF	Meta Object Facility	Model	General	Modelling	Yes	—	http://www.omg.org/
NAICS	North American Industry Classification System						

Table of Content Standardization Efforts

Acronym	Full Name	Semantic Depth	Domain Type	Focus	Meta Cont.	Bus. Cont.	Web address
NIIIP	National Industrial Information Infrastructure Protocols	Model	Process	Virtual Enterprise		Yes	http://www.niiis.org/
NIIIS	National Industrial Information Infrastructure	Model		EXPRESS models			http://www.niiis.org/
OBI	Open Buying on the Internet	Protocol	Process	B2B eCommerce		Yes	http://www.open-buy.com/
OCF	Online Catalog Format	Model	Product	Product Catalogs	No	Yes	http://www.martsoft.com/ocp/
OFX	Open Financial Exchange	Model	Product	Financial transactions	No	Yes	http://www.ofx.net/ofx/
POSC-CAE-SAR	Petrotechnical Open Software Corporation - CAE-SAR	Model	Product	Petroleum	No	Yes	http://www.posccae-sar.com/
PSL	Process Specification Language		Process		—	—	
RosettaNet	RosettaNet Catalog Interoperability Proposal	Dictionary	Product	Electronics	—	Yes	http://www.rosetta-net.org/
SWIFT	S.W.I.F.T.	Dictionary	Product	Finance transactions	No	Yes	http://www.swift.com/
UDEF	Universal Data Element Framework	Dictionary	General	Business	No	Yes	http://www.udef.com
UML Meta model	Universal Modelling Language Meta Model	Model	General		Yes	–	
UMLS			General	Healthcare	Yes	Yes	
UN/SPSC		Thesaurus	Product		—	Yes	http://www.spsc.org/
WfMC	Workflow Management Coalition	Model	Process	Workflow		Yes	http://www.wfmc.org/
xCBL	XML Common Business Library	Protocol	Process	Business documents handling	some	Yes	http://www.commerceone.com/xml/cbl/index.html
XML/EDI Repositories		Repository					www.xmledi.com/repository/

The M*-COMPLEX Approach to Enterprise Modeling, Engineering, and Integration

Giuseppe Berio and Antonio Di Leva

Dipartimento di Informatica, Università di Torino,
Corso Svizzera 185, I-10149 Torino, Italy
{berio, dileva}@di.unito.it

Abstract. A hot topic on the agenda of corporate management in order to improve the enterprise competitiveness factors concerns the operational, even continuos and rapid, transformation of business processes, support interoperability and integration with other business partners. The paper overviews through a simple case study concerning the definition of a virtual enterprise, the methodology M*-COMPLEX (extending authors previous word on M*-OBJECT) for design of process centered highly complex systems and its supporting tool M*-PROCESS.

1 Introduction

Business process concept has been recognized as the key element in inter and intra enterprise integration, management and virtual enterprise support. In fact, engineering business processes guarantees compliance with business requirements and overall system consistency and efficiency. Engineering in manufacturing is a systematic approach to the design of products and their manufacturing processes to address the QCD (Quality, Cost, Delay) challenge faced by most industrial companies and thus improve customer satisfaction [1]. The same applies to services for both design and availability to customers and users. Engineering business processes (and using standards as much as possible) of a virtual enterprise allows its components to behave as a unity just to covering special requests or needs for services or products while maximizing investment return and customer satisfaction.

In this paper, the M*-COMPLEX methodology is briefly presented together with M*-PROCESS, its tool for modeling and analysis of business processes. M*-COMPLEX is a structured framework which provides a step-by-step strategy ensuring consistent and correct results. It analyses functional, behavioral, information and organization aspects of the object organization, and it strongly enforces an event-driven process-based approach at all levels (the organizational level, the conceptual level, and the implementation level are considered) as opposed to traditional function-based approaches for analyzing and designing computer-supported integrated environments.

Some specific models and methodologies have been proposed in the literature for modeling and engineering information systems, such TSER [2]. Methodological approaches based on Object Oriented Analysis (see e.g. [3,4]) have been introduced to develop a conceptual model starting from user requirements. In our opinion, this often leads to a design in which end-users have to deal with rather abstract concepts like

S.W. Liddle, H.C. Mayr, B. Thalheim (Eds.): ER 2000 Workshop, LNCS 1921, pp. 7-18, 2000.
© Springer-Verlag Berlin Heidelberg 2000

classes, association types, attributes and value types. As a consequence, the resulting conceptual schema is not sufficiently transparent to end-users that have to validate it.

More generic approaches have been developed to model processes and information systems as well as organizations, such as CIMOSA [5], ARIS [6], or M*-OBJECT [7,8] which was originally developed for design and analysis of information systems in CIM environments. M*-OBJECT is organized as a three-stage approach according to the following phases:

1. **Organization Analysis**: this phase is used to analyze the business environment. Especially, functions and activities involved are identified, and flows of information are described.
2. **Conceptual Design**: this phase provides guidelines for building a formal specification of the information system, the *conceptual schema*, based on the PDN (Process and Data Net) model [8]. The conceptual schema is a detailed executable representation of static, dynamic, and behavioral enterprise features.
3. **Implementation Design**: this phase concerns logical and physical design of the database structure and architecture (possibly distributed or federated) to be implemented with the database management system selected for run-time operations.

M*-COMPLEX is the evolution of M*-OBJECT and is based on the assumption that a process-based (i.e. horizontal) approach must be adopted to model organizations for the purpose of integration rather than a function-based (i.e. vertical) approaches, because the aim of enterprise integration is to brake down organizational barriers and rationalize core process and information flows rather than creating "islands of automation". Basically, it differs from M*-OBJECT in the Organization Analysis phase because it adopts a new executable organization model which is supported by M*-PROCESS, a tool based upon the following prerequisites:

1. the environment should be highly interactive, with an icon-driven graphic editor to specify both functional and process nets; consistency rules must guarantee a coordinated development of different types of specifications;
2. process specifications should be executable by a discrete event simulation software that can be used to investigate different system functionality in order to bring into focus areas where implicit misunderstandings exist between users and designers.

M*-PROCESS has been developed on the top of a discrete event simulation engine, the Extend environment [9]. Process specifications are mainly based on IDEF3 language style (see e.g. [10]), however an ad-hoc formalism has been developed which specifically allows to describe behavioral interactions between running processes activities and resources. In that formal framework, behavior of a resource is represented through *Statecharts* [11] which main benefits are summarized in section 3.2.

This paper is organized as follows. Section 2 briefly illustrates the case study which will be used in the paper. In Section 3, the Organization Analysis phase of M*-COMPLEX is presented. Finally, Section 4 draws some conclusions.

2 The Case Study

Two enterprises that manufacture respectively products A and B have achieved an agreement to sell a new product C, assembling A and B. They need to manage the production process of C and their proper order processing procedures. The focus will be on the production process because the two enterprises would like to analyze the replication integrated with a flexible logistics on the site of A and on the site of B of assembling cell for C. In particular, the alternative assembling sites should allow:

– to reduce the impact of holidays for instance because enterprises A and B follow to distinct holidays calendars,
– to reduce the impact of failures with appropriate scheduling alternatives,
– to use a limited number of new physical resources.

The order management of A receives the customer orders for product C. This kind of orders splits into orders for A and orders for B. Because A sells C through its associated commercial structure, it needs to manage the new flow of money generated by the new product C. Moreover, money are transferred as soon as possible to the enterprise B after the enterprise A has been paid (even on partially paid orders).

3 The Organization Analysis Phase of M*-COMPLEX

The aim of the Organization Analysis phase is to define the scope of the project and to precisely analyze the current structure and state of the enterprise domain to be engineered or reengineered.

The analysis is mainly supported by an integrated "modeling language", i.e. a set of concepts which need to be used and shared both by analysts and business users. The integrated model consists of a functional model and a process model (also bringing a resource model), which will be illustrated in the case study. These models are based, respectively, on the IDEF0 and the IDEF3 models because the IDEF family of models, due to the simplicity and intuitive appeal of its graphical notations, represents the most widespread formalism for modeling and analysis of enterprises [10]. However, as pointed out in the introduction, IDEF3 is only an external and intuitive representation of processes because it only provides abstraction mechanisms for an underlying formal behavioral model which can be used as a basis for any representation suitable to be simulated through M*-PROCESS.

At the organization level, M*-COMPLEX views the world from two orthogonal points of view. First of all, an enterprise can be analyzed in terms of organization elements that can be classified as organization units (*units* for short) which control other units at a subordinated level, and so on. Units at the bottom level are called *work centers*. Units define areas of responsibilities and authorities and must be analyzed in order to identify their *functions*, i.e. things to be done and services to be provided. Top level functions are decomposed at different levels of detail, until the bottom level in which *activities* are carried on by work centers. From the other point of view, activities are executed by *resources*, processing or producing different *objects* (pure information or material objects). They are subject to scheduling or planning and can be coordinated into organization *processes*. Each process specifies the complex control flow between enterprise activities: it shows which activities should be performed at a time. Thus, an enterprise can be seen as a collection of concurrent *processes* which define the flow of actions and are triggered by stimuli called *events*.

The Organization Analysis phase of M*-COMPLEX is structured in two major steps: AS-IS Analysis and TO-BE Analysis.

The aim of the AS-IS step is to provide managers and engineers with an accurate model of the enterprise as it stands, out of which they can make a good assessment of its current status. It contains the following sequential tasks: *Structural Analysis, Functional Analysis, Process Reconstruction,* and *Validation.* Current problems are discovered during and listed into a report. Specifically, this is allowed by two kinds of analysis: qualitative (process and function schemata are inspected following detailed guidelines), and quantitative (the process schema is simulated with M*-PROCESS).

TO-BE analysis encompasses three major sequential tasks: *Diagnosis, Restructuring,* and *Validation,* which provides guidelines addressing the causes of the problems listed into the report issued from AS-IS step, and then try to solve them. During the Diagnosis task, problems are pointed out (issued from global or detailed diagnosis) along some causes; after that, possible solutions are suggested as global solutions or detailed solutions; last ones describe how and where to change different schema stated during the AS-IS step and represented through a *cause/solution matrix.* Decisions about which solutions should be followed are made during the Restructuring task. Finally, in the Validation task new solutions are checked against user requirements.

Both AS-IS and TO-BE steps may be iterated depending on the results from their respective validation tasks.

In order to properly apply M*-COMPLEX to virtual enterprise, during the first iteration of the TO-BE step, objectives of the virtual enterprise should be primarily taken into account for developing or changing processes, functions, organization units. Any further improvement of the virtual enterprise should be addressed by next TO-BE iterations, starting with problems issued from Validation tasks as described above.

3.1 The AS-IS Step

Due to space reason, the organization schema for the case study are not presented here.

Because A and B are both manufacturing industries, an example of their functional schema is shown in figure 1.

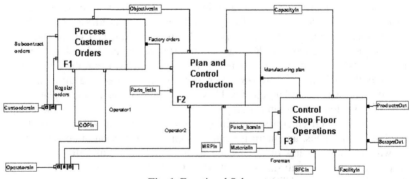

Fig. 1. Functional Schema

The Functional Analysis task is basically a top-down design phase, going from management down to production activities. The Process Reconstruction task is typically a bottom-up design phase in which elementary components (activities) are coordinated into process schema, and then aggregated to increase the clarity and expressiveness of the enterprise processes. The bottom-up strategy has been chosen to take full advantage of the results achieved in the Functional Analysis task. Indeed, if elementary components of the enterprise have already been analyzed, the bottom-up strategy is the simplest and most effective choice.

The co-ordination of activities can be carried out by taking into account their input and output objects, statuses, and events (*control items*) to discover causal relationships between them. Corresponding items must be recognized, and the process is obtained by merging corresponding input and output items of related activities. The methodology suggests a co-ordination strategy which starts from the process interface, consisting of items exchanged with the external world or other processes, and inserts into the schema the activities which are activated by the external items, then the routines activated by them, and so on. This "outside-in" strategy can be performed forward (input oriented) or backward (output oriented). The model is then enriched by proper identification of resources needed (e.g. agents, machines, applications).

For instance, in the two enterprises different parts are manufactured independently, thus the corresponding fragment of the production process is detailed in figure 2. The fragment represents a product P (in A or B) that firstly is transferred by a robot into buffers of a manufacturing cell, then worked, and finally stored somewhere.

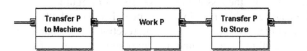

Fig. 2. Simple Working Process Schema

The co-ordination strategy produces a detailed specification of the process under analysis. Since a process can be composed by several activities (e.g. a hundred), it can be useful to clear up the layout by introducing sub-process blocks. The analyst should be careful to break the process at points with minimal connections between sub-processes. The fragment in figure 2 can be summarized at a more abstract level as the hierarchical block (sub-process) *Produce P*. In the figure 3, the whole manufacturing process for product P is illustrated.

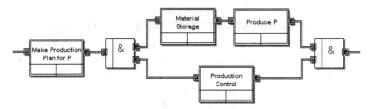

Fig. 3. Production Process Schema

In the case study, the Validation phase has the objective of validating the process representations. This task can be performed through simulation, process trace visualization, and object inspection, all supported by the M*-PROCESS tool. However, we do not show how it can be done because a similar approach will be followed in doing the TO-BE step: therefore, it will be deeply discussed in the remainder.

3.2 The TO-BE Step

The virtual organization redefines both the organization and functional schema of the two enterprises A and B. This simply means that:
- enterprise A sells the product C through its commercial structure,
- each enterprise A and B should be able to produce C,
- enterprise A accepts "orders" for C and therefore it has to manage money, transferring an appropriate quota to the enterprise B.

To achieve the objectives stated in section 2, production alternatives have to be represented. In fact, both enterprises A and B could be able to assemble C: this should allow to reduce the impact of failures in manufacturing cells, holidays, and shipping costs and time (e.g. producing C in the enterprise nearest some customers). Conceptually, the supplier relationship to produce C is dynamically established because sometimes A is a supplier of B and sometimes B is a supplier of A.

The virtual production process of the product C, named *Produce C*, containing all production alternatives is represented in figure 4. The bottom-up strategy is the best way for reusing activities and sub-processes already identified during the AS-IS step. However, whenever dealing with virtual enterprises, new activities and sub-processes are usually required (e.g., the Assembling sub-process in figure 4).

As it will be clarified in the remainder, the model supports both static and dynamic allocation of resources to activities. In fact, reactions of resources to unexpected events (e.g. re-scheduling, failure) may rise into some changes of the process status and dynamic re-allocation of resources to activities.

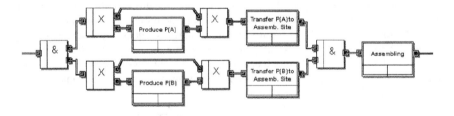

Fig. 4. The Working Process Schema For the Virtual Enterprise

In order to complete the representation of the new production process, resources have to be represented, possibly reusing resource schema already identified in the AS-IS step. In this case, possible resources are respectively
- the two Manufacturing Plants, MpA and MpB,
- the two Assembling Cells, AssA and AssB,
- the Carrier, a generic Track, to realize any transfer between the two plants,
- two Stores, each of them in a plant for stocking products A, B and C,
- two Robots, to realize any transfer inside each plant.

Figure 5 shown a representation by statecharts [11] of those principal resources: because most of them shown an equivalent behavior, they are generally represented with a final "X" just to remark that there are more than one individual resource with that behavior. The notation for statecharts transitions is standard as

Transition_Name: Events and/or Conditions: Functional Operations and/or Events

where only *Transition_Name* is mandatory. Bold ellipses are the default initial statuses. Complex statuses are named using capital types.

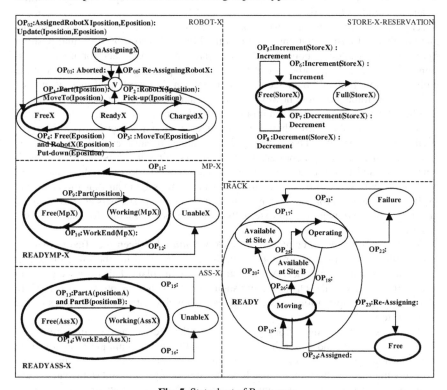

Fig. 5. Statechart of Resources

The STATEMATE semantics of statecharts [11] has been adopted in our framework. Main benefits of representing resources as statecharts are:

- resources can communicate between them and with the external environment through events,
- statuses of resources may represent pre-conditions, post-conditions of functional operations, and transitions represent what happens whenever some functional operations are executed,
- local variables parameterize functional operations, and events parameters,
- initial statuses can dynamically change depending on previous behavioral history, allowing some recovery,

- resources can behaviorally be aggregated through AND statuses to provide a new resource with a new integrated hierarchical behavior,
- operational behaviors of physical, human, informational resources are well modeled,
- reactions to events take effect iff resources are in some statuses in which some reaction to those events has been specified,
- compound transitions constitute an interesting way to represent behaviorally atomic complex transitions.

Specifically, the requirements space for the case study reuses robots as components that are able to transfer anything from an initial position *Iposition* to a final position *Eposition* (that can correspond to a set of physical positions). Each robot can be assigned by the appropriate process to cover any path depending on evaluation about the global situation. It should be noted that any robot can start working if there is at least a piece in the initial position: however, it cannot start if there is not any control object enabling the production of a given product, even if some piece is present in the initial location. This enabling flow should be triggered by an explicit event requiring production of some product. The same happens for plants because even if some piece is located for instance into the buffer of a working machine, it can be decided to work other pieces and therefore a choice between more control flows enabling machine's working activities has to be made. However, the plant assignment transitions are not explicitly modeled as for robots: any plant in its *Free(MpX)* status is able to follow a given *process plan*, which is an externally assigned local variable, for making some product.

An informational resource named *StoreXReservation* has been specified to manage any conflict between the two robots whenever they are used for the same objective, for instance to transfer products A to the store area in the plant. This resource gives an informational view of the physical store statuses but it is not the physical store. More specifically, the informational resource gives a view about the future statuses of the physical store.

Resources are sometimes strongly coupled, i.e. they behave as one single resource. For instance, a robot and a machine can be coupled to realize a unique behavior, from store to worked pieces without any particular interaction of processes. This means that a process interacts with robot and machine as a unique resource and statecharts represents robot and machine with a unique complex AND refined status (introducing a new level in the statecharts hierarchy).

In the formal framework the behavior of process fragment in figure 4 can be represented by a further refinement of IDEF3 specification into an appropriate Petri net class [12] (such as CPN or P/T, depending if either processes and resources should be identified and also if resource operations require parameters provided by processes [13]). That refinement takes into account resources of figure 5 and their functional operations. In fact, for each Petri net transition that needs a corresponding resource transition, a *synchronization relationship* is defined: however, there are both resources and processes transitions that are not part of any synchronization relationship, and they are referred as *non-synchronized*. Translating and refining any IDEF3 specification into Petri nets can be semi-automated by applying rules to connectors (i.e., &,X,O)[14] while activities, which job is usually resource dependent, will be translated following the behavioral structure of the involved resources. Firing two transitions participating in a synchronization relationship needs that both ones have to be enabled as respectively defined for Petri nets and statecharts. In the case study, the process fragment in figure 4 has been represented as the P/T net shown in figure 6 with the obvious correspondences.

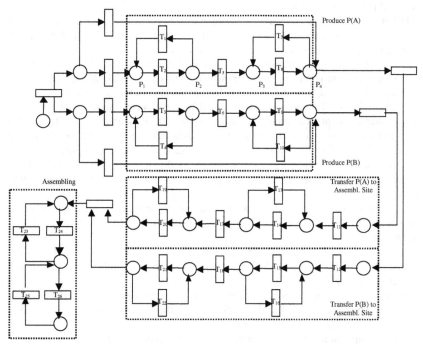

Fig. 6. P/T net Refining the Virtual Working Process in Figure 4

A possible synchronization relationship is as follow:

$\{(OP_1(RobotA),T_2), (OP_{02}(RobotA),T_1), (OP_9(MpA),T_5), (OP_1(RobotA),T_8),$
$(OP_{02}(RobotA),T_7), (OP_9(MpB),T_6), (OP_{17},T_{11}), (OP_1(RobotA),T_{14}),$
$(OP_{02}(RobotA),T_{13}), (OP_{25},T_{17}), (OP_1(RobotB),T_{20}), (OP_{02}(RobotB),T_{19}),$
$(OP_1(RobotA),T_{24}), (OP_{02}(RobotA),T_{23}), (OP_1(RobotB),T_{24}), (OP_{02}(RobotB),T_{23}),$
$(OP_{13}(AssA),T_{26}), (OP_{16}(AssA),T_{25}), (OP_{13}(AssB),T_{26}), (OP_{16}(AssB),T_{25}),$
$OP_1(RobotB),T_3), (OP_{02}(RobotB),T_4), (OP_9(MpB),T_6), (OP_1(RobotB),T_9),$
$(OP_{02}(RobotB),T_{10}), (OP_{25},T_{12}), (OP_1(RobotB),T_{15}), (OP_{02}(RobotB),T_{16}),$
$(OP_{17},T_{17}), (OP_1(RobotA),T_{21}), (OP_{02}(RobotA),T_{22})\}.$

The definition of the formal behavioral semantics of the processes/resources specification with statecharts and P/T nets can be found in [15]. The main benefits of this kind of specification style are listed hereinafter:

1. A synchronization relationship represents in a declarative style the control over processes and resources; in fact, both resources and processes can make decisions on which enabled transitions should be executed; therefore, failures, holidays, illness, special politics for resource substitution can easily be represented;

2. refinement of statecharts statuses can be used to provide more detailed resource specifications with or without impact on processes specifications;

3. while process behavior is well represented through Petri nets that allow to fold more process instances into a global specification, resource behavior should be represented in a unfolded manner; in fact, dealing with resources means to well represent constraints to be satisfied by each individual resource instance under concurrent and parallel accesses;

4. relationships between concurrent, cooperative processes can be represented through synchronization over shared resources, even in the case in which processes belong to distinct levels of enterprise management;

5. as opposed to the events treatment by statecharts, modeling processes by Petri nets allows to represent persistent events to which some reactions are always required.

Concerning the working process shown in figure 6, it should be noted that:

1. transitions T_1 and T_4 have to be represented because on one hand, a plant running with a particular process plan to produce a product D (neither A nor B) may be used to fire transition T_5 and that is not right (although this effect can be eliminated by specializing plant transition OP_9 on the process plan needed to produce a particular product while working a given piece); on the other hand, robots have to re-try any previously aborted transfer of a piece to be worked according to some process plan;

2. if the transfer of any piece to be worked has been completed, the plant working process of it can be delayed and that piece would not be restocked in its appropriate storing area;

3. there are non-synchronized transitions (the unnamed ones) that are used to make explicit control decisions at the same time resource independent and production objectives dependent;

4. the information resource *StoreXReservation* is not used here although, if needed, it can easily be synchronized with a transition in the P/T net specified before any other transition synchronized with robots;

5. the synchronization relationship states that a robot initiates a transfer from a store to machines iff it is enabled by the production process (couple (OP_1,T_2)); in fact, the working process concerns the production of one product A or B or C; however, another possibility is to consider the robot to be able to realize many transfers for a single order of many products A or B or C; in this last scenario, a more appropriate couple is (OP_{02},T_2) (analogous approach could be followed for other resources).

The formal framework based on Petri nets and statecharts integration (or interoperability) other than allowing a well founded and complete definition of complex behaviors, can profitably be used to precisely estimate some parameters required for simulation based on IDEF3 process specification. For instance, for the activity *Work A* (figure 2) it is easy to see that:

$$Time(Work\ A) = Time(T_5)+Time(P_3)$$

where $Time(T_5)$ is "inerithed" from transition OP_9 of MpA and $Time(P_3)$ is essentially the time needed by the resource MpA for returning in its Free(MpA) status. By simulating with M*-PROCESS, aggregated and interesting parameters of the process *Produce C* can effectively be evaluated and further used to simulate the global manufacturing process.

The global virtual manufacturing process (figure 7) generates two production plans, showing the two possible alternatives for producing C: A acting as supplier of B and B acting as a supplier of A. Therefore, these two production plans (PP) are generated (activities *PP for C on A* and *PP for C on B*), evaluated and then a choice is made between them (activity *Make Choice*). However, the rejected production plan may continuously be updated and conserved as a future alternative. Going further, the manufacturing process enables respectively the material (or product) storage and the production of C and, in parallel, both the production and logistic control. The logistic control is important to control the running tracks, and to solve any problem arising during both products and material shipping.

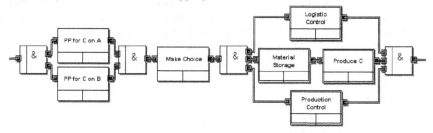

Fig. 7. The Production Process for the Virtual Enterprise

The order payment management for product C (figure 8) focuses on production of payment orders, payment verification and money transfer from the enterprise A to the enterprise B according to their agreement. This process has an administrative nature and its principal resources are typically humans: therefore, illness or holidays of employees are typical kinds of resource behaviors which impacts on the process behavior should be evaluated.

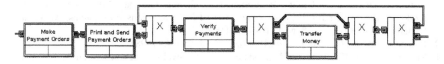

Fig. 8. Treatment of Payment Orders of the Virtual Enterprise

The first iteration of the TO-BE step ends. Therefore, the Validation task starts and it should establish if the provided solution for implementing the virtual organization has been fully achieved. If not, a new iteration of TO-BE step to improve current solution will take place.

Remark. The organization analysis of the virtual enterprise provides the set of requirements for starting more detailed analysis and implementation. M*-OBJECT suggests to represent, through PDN, conceptual objects and their classes as well as the impact of processes on object statuses. For instance, from the process schema in figure 8, a PDN schema can further be developed. In this specific case, each process activity can be refined into a corresponding PDN transition which acts on both informational resources and other data objects, and codes how to transform this information. In this manner, the impact of the order payment process on conceptual data can formally be stated and it can be used to provide, for instance, a process implementation driven by data changes (as in document or database management systems).

4 Conclusion

An approach for analyzing, modeling and implementing business processes for enterprise engineering, modeling and integration has been presented using the M*-COMPLEX methodology. In this paper, it has also been shown how M*-COMPLEX can be used to define engineering processes using a process-based (or horizontal) approach, as opposed to traditional function-based (or vertical) approaches. With the tool M*-PROCESS, this approach effectively provides the basis for designing "what-if" scenarios, i.e., reliable, virtual approximations of how changing process will impact an enterprise before it invests resources in implementation. Furthermore, developed schema can also be used to really implement processes both on top of some workflow management systems and as information system evolution.

References

1. Karagiannis, D., (ed.):   Special Issue in Business Process Reengineering. In SIGOIS Bulletin, Vol. 16, N. 1 (1995)
2. Hsu, C., Rattner, L.: Information modelling for computerized manufacturing. In IEEE Transactions on Systems, Man, and Cybernetics, Vol. 20, N. 4 (1990) 758-776
3. Rumbaugh, J., et al.: Object Oriented Modeling and Design. Prentice Hall (1991)
4. Rumbaugh, J., et al.: The Unified Modeling Language User Guide. Addison-Wesley (1999)
5. ESPRIT Consortium AMICE (eds.): CIMOSA, Open System Architecture for CIM. 2nd edition, Springer –Verlag (1993)
6. Sheer, A.W.: ARIS – Business Process Frameworks. Springer Verlag (1999)
7. Berio, G., Di Leva, A., Giolito, P., Vernadat, F.: The M*-OBJECT methodology for information system design in CIM environments. In IEEE Transactions on Systems, Man, and Cybernetics, Vol. 25, N.1 (1995) 68-85
8. Berio, G., Di Leva, A., Giolito, P., Vernadat, F.: Process and Data Nets: The conceptual model of the M*-OBJECT methodology. In IEEE Transactions on Systems, Man, and Cybernetics, Part B, Vol. 28, N. 1 (1999) 104-114
9. Extend 4.1 User's Manual. ImagineThat Inc., San Jose, CA (1998)
10. Mayer, R.J.: The IDEF Suite of Methods for System Development and Evolution. Technical Report, Air Force Human Resources Laboratory, Wright-Patterson Air Force Base, Ohio 45433-6503 (1991)
11. Harel, D., Naamad, A.: The STATEMATE Semantics of Statecharts. In ACM Transactions on Software Engineering and Methodology, Vol.4, N.5 (1996) 293-333
12. Brauer, W., Reisig, W., Rozemberg, G., (eds.): Petri Nets: Central Models and their Properties. Advances in Petri Nets, Part I and II. Lecture Notes in Computer Science, Vols. 254, 255. Springer Verlag, Berlin (1987)
13. Berio, G., Vernadat, F.: Une Méthode de Spécification du Comportement des Systèmes Réactifs de Production. In Proceedings 2ème Conférence Francophone de Modélisation et Simulation (MOSIM'99) (1999) 185-190
14. Vernadat, F.: Enterprise Modeling and Integration: Principles and Applications. Chapman & Hall, London (1996)
15. Berio, G., Vernadat, F.: Formal Foundations for a Process/Resource Approach in Manufacturing Systems Behaviour Modelling. In Proceedings 14th IFAC World Congress (IFAC'99) (1999) 181-186

Conceptual Design of Electronic Product Catalogs Using Object-Oriented Hypermedia Modeling Techniques

Cristina Cachero[1*], Jaime Gómez[1], and Oscar Pastor[2]

[1] Departamento de Lenguajes y Sistemas Informáticos
Universidad de Alicante. SPAIN
{ccachero,jgomez}@dlsi.ua.es
[2] Departamento de Sistemas Informáticos y Computación
Universidad Politécnica de Valencia. SPAIN
opastor@dsic.upv.es

Abstract The application of conceptual models that assure both the consistency and usability of Electronic Product Catalogs (EPC's) is a main concern in the e-commerce community, mainly due to the impact a correct design of such interfaces has on the final e-store sales figures. This article describes how the OO-$\mathcal{H}$Method web-interface modeling approach can be successfully applied to the design of Business to Consumer (B2C) application interfaces. In order to address the specific domain semantics, OO-$\mathcal{H}$Method adds a new concept called 'dimension attribute', which is extracted from the datawarehouse field, and defines a comparison function applicable to these attributes. Also, a set of e-commerce Interface Usability Patterns that address some design problems likely to appear on the different store components and a materialization strategy are introduced to complete our proposal. The use of patterns to capture the semantics specifically related to e-commerce applications, and to manage the knowledge captured in the catalog structure, guarantees the reuse of techniques coming from different fields and proven useful for the increment of sales on internet.

1 Introduction

In the last few years the growth of the Internet market has supposed a main challenge for the enterprises, that have had to quickly adapt both its technology and its business processes in order to get its share in such potential benefits. In order to facilitate this evolution, the research community has been compelled to work on the several orthogonal aspects involved in this process, and specially in the design and adoption of new e-commerce practices. Those aspects include concepts such as technology availability, interoperability, security, control of electronic transactions, etc., which apply to every kind of e-business application. In

---

* This article has been written with the sponsorship of the Conselleria de Cultura, Educació i Ciència de la Comunitat Valenciana

S.W. Liddle, H.C. Mayr, B. Thalheim (Eds.): ER 2000 Workshop, LNCS 1921, pp. 19–30, 2000.

this article, however, we will center on Business to Consumer (B2C) and Consumer to Consumer (C2C) applications, where the interface design process and usability features are especially relevant: the (above mentioned) need of integration with the general enterprise business requires a multidisciplinary development process, and the large target audience and customer desired behaviour (coming back, purchasing our items) requires meaningful metaphors and well designed navigation paths. Supporting this assertion, performed research [10] has concluded that improving browsing and navigation capabilities, and especially product lists, is more relevant than visually striking store fronts and can generate significantly higher traffic and sales per store. These are precisely two of the main characteristics defined by means of an Electronic Product Catalog (EPC) [13,9].

EPC's can be seen as Web Information Systems [3] that, in addition to laying emphasis on the presentation of products/services, contain some standard functionality regarding navigation, searching, selection and ordering of products [7, 8]. They materialize the paper catalog metaphor, but, due to the fact that they provide a much more powerful source of information on products, and that purchase decisions are usually made on this information, they are considerably more effective. In order to better attain the customer attention and favor a purchase behaviour, EPC's tend to be highly interactive and dynamic. Furthermore, as the consumer is highly sensible to errors (out-of-date, imprecise information and/or communication errors), they need to be specially carefully designed. Both the information quality (accurate and up-to-date information, availability of items, special offers...), its organization and the facilities provided to improve the buyer experience (such as comparison facilities) influence the success of an e-commerce application.

In spite of these facts there is, as far as we know, no clear proposal that specifically deals with the idiosyncrasy of such applications. Efforts are being made among enterprises in conjunction with their immediate trading partners, and the danger is that this relatively narrow focus limits the return on investment possible from each of these initiatives. E-commerce needs a global environment, and the application of a coherent conceptual model can help to achieve it, creating at the same time a common framework for further discussion. Moreover, using a conceptual model the application becomes less sensible to the technological changes which are constantly appearing in this environment. Traditional software engineering techniques, even if they can suffice for a first approach, do not encourage the designer to think about and capture the specific semantics for EPC's. An enrichment of existing modeling techniques would allow to attain a faster, more reliable and thus cheaper development process, together with a better turnover for the enterprises. Following this trend, in this paper we are extending our proposal, known as OO-$\mathcal{H}$Method [6], to fulfill this purpose. Furthermore, we will show how some well known e-commerce interface usability patterns can be applied to improve the effectiveness of EPC's.

The remainder of the article is structured as follows: section 2 provides a brief introduction to the OO-$\mathcal{H}$Method web-modeling approach, describing the diagrams that are used to capture the relevant properties of a web interface. Section 3 extends our proposal in order to capture the concepts related to EPC's. Some of these concepts have been borrowed from the datawarehouse field. That is the case of the dimension attributes and the compare operations associated to them. Some of the more relevant EPC patterns, gathered in an OO-$\mathcal{H}$Method Pattern Catalog and used to improve the EPC interface, are described in section 4, together with the Abstract Presentation Diagram on which they are applied. Finally, section 5 presents the conclusions and further work.

2   Basic Features of OO-$\mathcal{H}$Method

OO-$\mathcal{H}$Method is a generic model for the semantic structure of web interfaces. It captures the relevant properties of a web interface by means of two models: the navigation model and the presentation model. The navigation model is captured by means of the Navigation Access Diagram (NAD). It captures the information which each potential web user (agent) can access and the navigation paths from one information view to another. The presentation model is specified by means of the Abstract Presentation Diagram (APD) and provides a set of default values which might be assumed for all the presentation features in order for a web interface to be automatically generated without further work. To do it, OO-$\mathcal{H}$Method defines the mapping from elements in the NAD to elements in the APD, as well as some possible ways of refinement.

2.1   The Navigation Access Diagram (NAD)

The NAD provides four major conceptual modeling constructs to specify navigational user requirements: (1) Navigation Classes (NC), (2) Navigation Targets (NT), (3) Navigation Links (NL) and (4) Collections (C). Navigation Classes are domain classes whose attributes and methods have been filtered and enriched in order to better accommodate the specific features of hypertext. Navigation Targets group the elements of the model that collaborate in the coverage of a certain user navigation requirement. Navigation Links define the navigation paths through the information. Finally, Collections are hierarchical structures defined on navigation classes which define both the traversal behaviour of each one of the links grouped by the collection and the set of objects on which this collection will apply.

2.2   The Abstract Presentation Diagram (APD)

The APD adopts a template approach [1,4,5,11] for the specification of not only the visual appearance but also the page structure of the web. As far as we know, there are five main aspects that should be specifically considered when generating a web interface, namely, the information to be shown, the layout, the interaction processes, the client functionality and the possibility of multiple simultaneous views. These aspects are gathered in five template types, each one with its corresponding Document Type Definition (DTD) associated. The template types are, namely: tStruct, tStyle, tForm, tFunction and tWindow. These diagrams will be further discussed in the rest of the paper.

3   Extending OO-$\mathcal{H}$Method to Cover EPC's Semantics

We have already stated that one of the main differentiating characteristics of EPC's is their specially flexible browsing and navigating capabilities. This flexible navigation can be compared to the one we can find in the OLAP domain [17]. In the same way the OLAP user manipulates information in order to visualize the data in a format that allows him to make a business decision, the EPC customer must have the possibility to manipulate the product information contained in the catalog in order to make a buying decision. For a more general perspective of the approach, a small example is going to be employed all along the paper: the online catalog of a music sample store called 'ShopAtOnce'. As a basic explanation (for reasons of brevity) it is assumed that the products (CD's, singles) are classified into categories (pop-rock, jazz, classic, latin music...) through which the user can navigate. The store also sells complementary products: t-shirts with the logos of the different groups, posters etc. The system implements the Shopping Cart Metaphor, directly available from any part of the store. The basic navigation requirements for our sample application have been defined as follows:

- On entering the site, the customer has access to the Hot Deals of the store. S/he will see the name, short description, retail price, saving (compared to the normal price) and a small image of the product.
- Also, the EPC should contain a search capability. There the customer will be able to choose whether s/he wants to search the CD's, either by title or by singer.
- The customer will also be able to navigate through the different categories of CD's. As a restriction, the only products that can be scanned are those that the company has in stock. Whenever the customer is visualizing a product, s/he should be able both to find complementary products (in our case, posters of the selected group, singles etc), add the actual product to the shopping cart and perform comparisons with other products in the catalog.
- At any time, the customer will be able to delete an item from the shopping cart, modify the quantity, print the invoice with the total amount (shipping and tax included) and/or perform the checkout process.

Figure 1 illustrates the NAD of our sample application for a standard customer profile. In this diagram we can identify four main Navigation Targets (NT) which cover the Hot Deals, the Search Catalog and the Scan Products requirements, as well as the Shopping Cart functionality (left out of our model due to the fact that it is a well-known component with a clearly defined behaviour). The NT define the main tasks the consumer can perform when browsing through the EPC. The entry point to the application gathers these four NT. 'Show all'

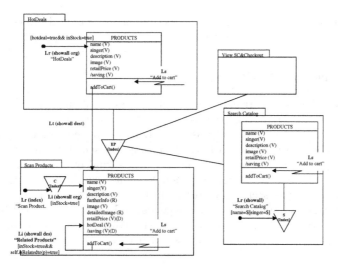

Figure 1. *NAD of the ShopAtOnce Electronic Product Catalog*

navigation patterns determine that all the Visible Attributes (V) related to the target objects for a given link or collection will be shown. Whether this visualization is performed on the origin page or by means of a new page depends on the 'org/dest' attribute associated to the navigation pattern. Similarly, the index pattern indicates that references to every object involved in a given navigation requirement are going to be shown together (although possibly paginated), and clicking on any of them will generate a new page with all the visible items of information available. Furthermore, Visible Attributes (V) determine the set of information relevant for the customer. Further details can be obtained by means of a 'More Info' link, generated when the designer characterizes one or more Referenced Attributes (R) on a given class. Restrictions might be applied both to the target population of links and collections, by means of filters. As an example, the 'in Stock=true' filter specifies that only products in stock will be shown as purchase possibilities. On the contrary, filters associated to Service Links (arrow symbols) determine attributes necessary for launching a given class method. In our example we have only one service available: add a product to the shopping cart. The additional characteristic of 'searching complementary products' is modeled by means of an Internal Link and a filter associated that checks for a given association among products in the underlying conceptual schema. Reviewing the requirements, the only operations left are those related to product comparison. This kind of operations, required by the customer and inherent to the buying process, are different from the traditional navigation behaviour the model captures among classes. It involves a new kind of internal, usually bifocal navigation which will be explained later and depends on the semantics of the attributes. Thus, we have introduced a new concept, that of 'dimension attributes', which together with the EPC Comparison Pattern defines the way the customer can get reassured in his/her purchase decision. Both concepts will be developed in the following sections.

3.1   Dimension Attributes

We define dimension attributes (D-Attributes) as attributes on which it is possible to define a comparison process. They define a new kind of navigation paths inside the Product conceptual class. D-Attributes are represented by means of a (D) symbol close to the attribute. An example of such attributes can be seen in Figure 1. Dimension attributes qualify a new kind of object collection that remains implicit in the model. An example about how this comparison process is performed can be seen in Figure 5. On this new kind of navigation path, the possible target set of objects always depends on the values of the attributes considered relevant for comparison (dimension attributes) on the actual product. Also, the set of applicable operators will depend on the attribute type.

3.2   Navigation on Dimension Attributes

We have already stated that both the customer browsing an EPC and a marketing or sales analyst browsing an OLAP application have a similar aim: extracting objective information on which to base a decision (either a purchase or a management decision). In fact EPC's can be seen as the front-end from which to gather all the information needed to apply OLAP operations. Aggregating and disaggregating products can be seen as navigating either through class hierarchies, class aggregations or even hierarchical collections. However, unlike on OLAP roll-up and drill-down operations, this aggregation/disaggregation means a change not only in the layout but also in the set of objects being visualized. Up to now we have not encountered the necessity of including a new kind of link that just changes visualization without affecting target population of the link. However we do not discard the possibility of performing such enrichment in the future. Other OLAP operations can be defined (although statically) by means of a different visibility scope for attributes (slice) or filters on links and collections (sub-population that can be thought as a sub-cube to be worked on). However, there have been identified some internal navigation needs, namely the comparison need, which is similar in both kinds of applications. In OLAP systems comparisons are usually visually performed, together with an interpretation of the data. As the visualization is usually made by means of tables, this comparison way appears appropriate. However, when trying to buy a product, we are given lots of additional information, and so the comparison becomes more confusing. By letting the user know the semantically relevant differentiating attributes (dimension attributes), together with its corresponding comparison operations, we are facilitating the purchase decision adoption and thus improving the interface usability. Once the navigation facilities are clearly defined through the classes and attributes of the system, it is time to go one step further and generate an abstract interface that gathers the main features of the EPC.

4   Generating the EPC Front-End

Following the navigation semantics captured in the NAD diagram, a default APD diagram can be generated and, from there, an EPC interface prototype can be derived in an automated way. However, there are some characteristics that can contribute to the improvement of the EPC usability features. Although the rapid development of the web business environment makes difficult the identification of best business practices, some of these features have already been identified in the e-commerce literature [14] and have thus been integrated in the Pattern Catalog [2] already available in OO-$\mathcal{H}$Method. Furthermore, in the e-commerce environment, companies cannot wait, as time means possibly a loss in the market share. So, we consider that, by providing this set of EPC interface patterns, we are helping the designer to identify what works without having to 'try and err', and so they are likely to improve the general effectiveness of the application while shortening the development period. In the next section we are going to show the main EPC patterns applied to our sample application. In OO-$\mathcal{H}$Method there is a set of rules that directly apply to the NAD concepts to generate a default diagram containing a set of Abstract Pages belonging to one of the five possible page templates defined: tStruct, tForm, tWindow, tFunction and tStyle. These rules fall out of the scope of this article.

4.1   EPC Patterns and APD Refinement

The EPC Pattern chapter is, as the rest of the OO-$\mathcal{H}$Method Pattern Catalog, user centered, that is, it provides additional mechanisms to fulfill the user requirements. The EPC patterns captured in the catalog offer alternative solutions to well-known product navigation and visualization problems. Due to the lack of space, in this article we will just show the patterns that have been applied to the default APD generated from the NAD in order to refine it. Next we are providing an informal description of such patterns:

– Comparison Pattern [12]: It gives information about the buying context of the customer. The problem it tries to solve is the lack of confidence of the customer on being getting the best deal when scanning traditional product Catalogs. There, the products are usually shown one by one, without further references among them. To avoid this discomfort feeling and improve the usability of the EPC we need a mechanism to provide comparison information. The solution is to implement a mechanism that allows the customer to compare, at any time, the product s/he is considering to buy according to a set of attribute dimensions previously defined, and allow further navigation through this comparison. There are several comparison possibilities: we can distinguish between bifocal comparison (default implementation), which implies that the product on which the comparison is initiated is never lost, and unifocal comparison, which provides new dynamic navigation collections that become independent of the original product.

– Shopping Cart Visibility Pattern [15]: It provides direct access to the state of the purchase from any page in the EPC. The problem it tries to avoid is the customer not being confortable when adding items to the cart just in case s/he decides not to buy it afterwards. There have been found some e-stores where you could not see what you had bought (and all the data associated, such as shipping costs, total amount etc) until you performed the checkout process, and even then you could just either accept the whole purchase or roll back the whole transaction. The solution is implementing the Shopping Cart as a component reachable from any navigation requirement, either by means of a global menu (default implementation) or by means of links from every page of the EPC.

– Payment Pattern [13]: It provides alternative payment methods in case the customer is not confident on the transaction security. The solution is always provide the possibility to contact a person (either by telephone, by fax etc) and finish the transaction over that other media. The easiest way of implementation, and thus the one by default, is giving on every page the e-mail and telephone of the sales office so that the customer has this information always available. It can also be enriched by means of a help page dealing with the possible payment policies of the store.

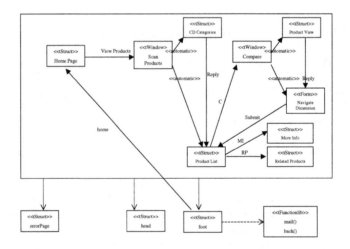

Figure 2. *Partial APD of the ShopAtOnce Electronic Product Catalog*

Just note that, apart from these patterns, we have also applied other more general ones, such as the Location Pattern [16] by means of including a header and footer on every page, and the Confirmation Pattern in order to confirm the goods purchase. Figure 2 shows a partial view of the default APD diagram, which has been enriched with the patterns defined above. This diagram describes the Abstract Page Structure of the Scan Product Navigation Targets.

As previously stated, the modeling constructs of the APD consist of a set of templates expressed in XML. From these templates, a materialization strategy allows the automatic generation of the EPC interface in the desired environment. One possible implementation, based on HTML technology, will be described in the following section.

4.2   Generated Interface

Figures 3 to 5 illustrate the prototype generated from the refined APD. The process is as follows: first, the generator tool looks for the Entry Point Collection of the application. In our example the navigation mode is set to 'index', and the chosen interaction technique (whose configuration process goes beyond the scope of this article) is a text menu bar which will remain visible on the top of every screen (see Figure 3).

Figure 3. ShopAtOnce entry point   **Figure 4.** ShopAtOnce Scan Products

Moreover, we can observe that the requirement link of the Navigation Target 'HotDeals' has a 'show all in origin' navigation pattern associated. This fact causes the home page to be enriched with the result of selecting all the CD's categorized as Hot Deals (which is specified by means of the corresponding filter). Besides, as the ShopAtOnce store has decided not to offer anything that is out-of-stock (as stated in the navigation requirements), the condition 'inStock=true' guarantees that all the products listed conform the restriction. Also, the existence of a service link pointing at the 'addToCart' function causes a Cart symbol to appear on every product. Also, we can observe how the information showed is that categorized as always Visible (V) in the NAD diagram. The existence of a link to a more detailed view of a given product is justified by the fact that there is a Traversal Link from the HotDeals NT to the Scan Products NT. As this link is given no name, the identifying attribute of the hotdeal is considered as the linking text.

The generation of the 'Scan Products' screen (see Figure 4) follows a similar process: as there is a Classifying Collection with a 'show all in origin' navigation

Figure 5. ShopAtOnce Comparison Process and Search Facility

pattern associated, this collection will show its structure directly on the page. However, from the collection to the navigation class there is also a 'show all in origin' navigation filter associated. This causes the generation tool to create a tFrame page that makes possible the two simultaneous views. Initially, the first option (show ALL products) is activated. The 'show all' navigation pattern has an 'items per page' attribute associated, which is in this case set to six. This attribute tries to avoid both overwhelming the customer with too much information on a single page and having to scroll through it, and causes the 'next' and 'previous' arrows to appear on the top of the screen. Next to the identifying attribute of the product there are three options: 'C' (for Compare), although we could have designed an icon for a more intuitive navigation) brings the customer to the Comparison screen (see Figure 5). 'MI' would give the user a more detailed view of the class. Although not shown in the storyboard, this page would show all the referenced (R) attributes of the product together with the visible ones, the option of adding the item to the cart and a back button to the previous screen. Last, but not least, the 'RP' link would bring the user to a new screen showing the Related Products (t-shirts of the group, posters and whatever item the customer might be interested in). This link comes from the Li showed in the NAD, and thus the navigation pattern associated also specifies its layout. In order to define the Comparison screen we have to again gather information from the NAD (see Figure 1): we can see the dimension attributes specified on this target. Those are the retailPrice and the percentage saved by the customer that buys at our store over the official price. As stated above, the customer will be able to compare products based on these articles. In order to implement this feature, we have followed a 'bifocal approach': the generation tool automatically generates a reduced view of the main product, and on the right side of the screen it opens a new view which starts by a tForm page. On this tForm we specify both the attributes on which we want to compare, and the operators we want to apply. The operators will depend on the type of attribute. A number attribute will have the options 'greater than', 'equal', 'distinct' or 'lower than'. An string attribute will provide the 'similar to' operation and will allow the use of masks. The sample screen for this example can be seen in Figure

5. Once the user has entered the desired values, the system will show, together with the product being compared, an index page containing all the product that match the comparing conditions. The last requirement is that of searching a product either by singer or by CD name. The resulting screen is, again, of the type tForm. When introducing either of the two fields (or both), the resulting page will give you an index page of type tStruct, containing the objects that match your query with all the specified attributes (showall pattern).

5   Conclusions and Further Work

An EPC has a well defined structure, which facilitates the automatization of many of the processes involved in the construction of e-commerce sites. As an example we could give the Site Server (www.microsoft.com/siteserver) from Microsoft or the Intershop web site (www.intershop.com), where we can find e-commerce general software solutions. However, in order to provide the customer with semantically relevant navigation and facilities through the catalog, the specific characteristics of the product must be taken into account. As an example, we can look for clothes color-matching to the one we are buying, but we rarely will look for a car color-matching to the one we are visualizing. Similarly, a tool that helps the user choose the correct t-shirt size is not usually provided by a general tool. On the other hand, existing OO modeling approaches are usually far too general, and thus obligue to model again and again the same recurrent concepts common to every EPC application, making the modeling experience rather tiring. Moreover, modeling approaches without possibilities of code generation lead to too long development periods, which cannot be afforded by enterprises competing in an e-commerce environment.In this context, the main contributions of our work could be summarized as follows:

- Application and extension of the OO-$\mathcal{H}$Method modeling approach to cover the idiosyncrasy of EPC's.
- The description of the way for obtaining a natural integration of the main OLAP operations (slice, dice, roll-up, drill-down) in an OO-$\mathcal{H}$Method model.
- The extraction of some usability concepts and its capture in the model by means of patterns with its corresponding APD materialization.

We are aware that the code generated up to now is much simpler than that provided by some commercial wizards. Even if has been proven that appalling interfaces do not seem to have much impact on the purchase attitude, we are currently working on the improvement of this design appearance by constructing a new tool that, departing from the Abstract Pages, give the designer more control both on the different interaction techniques and on the layout of the elements. Also, we are interested in the integration of the catalogs generated with OO-$\mathcal{H}$Method with other EPC systems. There are several alternatives, one of which could be the integration with the eCo System, a framework developed at CommerceNet(www.commercenet.com). As we are already generating abstract pages defined in XML, we think that, whatever the approach finally taken, the translation process will be rather straightforward. The interaction among systems is necessary in order to interact with software agents and search engines.

References

1. P. Atzeni, G. Mecca, and P. Merialdo. Design and Maintenance of Data-Intensive Web Sites. In *Advances in Database Technology - EDBT'98*, pages 436–449, 03 1998.
2. C. Cachero. The OO-$\mathcal{H}$Method Pattern Catalog. Technical report, Universidad de Alicante, 12 1999.
3. S. Ceri, P. Fraternali, and S. Paraboschi. Design Principles for Data-Intensive Web Sites. *SIGMOD Record*, 28:84–89, 03 1999.
4. F. M. Fernández, D. Florescu, J. Kang, A. Levy, and D. Suciu. Catching the Boat with Strudel: Experiences with a Web-Site Management System. In *Proceedings of ACM SIGMOD International conference on Management of data*, pages 414–425, 10 1998.
5. P. Fraternali and P. Paolini. A Conceptual Model and a Tool Environment for Developing more Scalable, Dynamic, and Customizable Web Applications. In *Advances in Database Technology - EDBT'98*, pages 421–435, 1998.
6. J. Gómez, C. Cachero, and O. Pastor. Extending a Conceptual Modelling Approach to Web Application Design. In *CAiSE '00. 12th International Conference on Advanced Information Systems. Lecture Notes in Computer Science*, volume 1789, pages 79–93. Springer-Verlag., 06 2000.
7. N. Koch and L. Mardiel. State of the Art and Classification of Electronic Product Catalogues on CD-ROM. *Electronic Markets*, 7(3):16–21, 1997.
8. N. Koch and A. Turk. Towards a Methodical Development of Electronic Catalogues. *Electronic Markets*, 7(3):28–31, 1997.
9. D. M. Lincke and B. Schmid. Mediating Electronic Product Catalogs. *Communication of the ACM*, 41(7):86–88, 1998.
10. G. L. Lohse and P. Spiller. Quantifying the Effect of User Interface Design Features on Cyberstore Traffic and Sales. In *CHI'98 Proceedings*, pages 211–218. ACM, 1991.
11. G. Mecca, P. Merialdo, P. Atzeni, and V. Crescenzi. The ARANEUS Guide to Web-Site Development. Technical report, Universidad de Roma, 03 1999.
12. Maria Milosavljevic and Jon Oberlander. Dynamic Hypertext Catalogues: helping users to help themselves. In *Proceedings of second ACM conference on HYPERTEXT'98*, pages 123–131, 1998.
13. J. W. Palmer. Retailing on the WWW: The Use of Electronic Product Catalogs. *Electronic Markets*, 7(3):6–9, 1997.
14. J. A. Rohn. Creating Usable E-Commerce Sites. *StandardView*, 6(3):110–115, 1998.
15. J. A. Rohn. Electronic Shopping. *Communication of the ACM*, 41(7):81–88, 1998.
16. G. Rossi, D. Schwabe, and A. Garrido. Design Reuse in Hypermedia Applications Development. In *Proceedings of the eight ACM conference on HYPERTEXT '97*, pages 57–66, 1997.
17. J. Trujillo, M. Palomar, and J. Gómez. Detecting patterns and OLAP operations in the GOLD model. In *Proceedings of the ACM 2nd International Workshop on Data warehousing and OLAP*, pages 48–53, Kansas City, Missouri, USA, November 1999.

Generic Linear Business Process Modeling

Guido Schimm

OFFIS, Escherweg 2, D-26121 Oldenburg
schimm@offis.de

Abstract. Visual-oriented process modeling concepts and languages used today often lead to modeling of "spaghetti-processes". Thus, the process models are not applicable effectively in distributed, partially automated and dynamic scenarios. They also support permanent adaptation and composition of process models insufficiently. The approach of Generic Linear Business Process Modeling describes a meta model that aims to avoid these inadequacy. According to the concept of linear programming, nested logic blocks are applied to model the process flow. Additionally, by using generic blocks composition and amplifying of process models on different levels of detail is achieved. Further the approach outlines the visualization of generic linear process models and an XML based format in order to exchange the models in a networked business environment.

Introduction

Process models are used for planning, designing, automatically supporting, monitoring, and recording business processes. These models represent the knowledge about the execution of business processes and act as schemata for storing the instances of processes or cases. This knowledge has to be provided in an easy to use meta model which ensures inherent consistency and adaptability. The emerging network economy enforces a number of requirements doing business process modeling: adaptability, support of composition and amplifying, ability of automatic execution/interpretation and exchangeability. These requirements are insufficiently covered by the state of the art of business process modeling.

As the business permanently changes today the process models need to be changed, too. The ability to respond effectively to changes ranges from ad-hoc modifications according to a single case up to evolutionary changes of many processes altogether [1]. Therefore the flexibility by selection provided by most modeling approaches has to be complemented with the flexibility by adaptation [5].

Business processes often involve a number of participants. This is especially true for business processes of enterprise networks. Here a whole process consists of various processes of different participants. In order to combine the partial processes flexibly, process models should be based on a meta model, which supports dynamic assembling of business processes in distributed environments. Also, it is evident that exchangeability of process models is required in a distributed environment.

Often a process starts without prearranging all process steps and detailed activities that have to be executed in order to reach the final process goal. Only a rough plan or scheme is available. While the process goes on the next activities are determined

S.W. Liddle, H.C. Mayr, B. Thalheim (Eds.): ER 2000 Workshop, LNCS 1921, pp. 31-39, 2000.
© Springer-Verlag Berlin Heidelberg 2000

successively. In this case a process model has to support its permanent amplifying even if the process is already running.

Deployment of new technologies, in particular internet technologies, is leading to new forms of business organization. If these business forms are to achieve competitive efficiency, partial automation of business processes is required. Today, most process models are fine for execution of rigid, less complex business processes with a high repetition frequency. But in a customer oriented business such processes are rather rare. In fact, flexible and dynamic process models are needed. These should be supported by computers reliably. Here, the diagram languages used today for business process modeling purposes often fail partially.

On the one hand, the diagram languages have the advantage that they visualize process models vivid. This advantage ultimately leads to their widespread distribution and therefore should be adopted by new modelling approaches. On the other hand, these languages often suffer from a lack of precise syntax and semantic definitions [6]. This often leads to process models that becomes inapplicable for automatic execution by computer systems in practice. Additionally, the important ability to respond effectively to changes is hindered. The approach on Generic Linear Business Process Modeling tries to avoid these deficiencies and is intended as an improvement in business process modeling according to the requirements of the network economy. It subsumes and combines existing concepts and techniques.

Building Blocks of Linear Process Modeling

The diagram languages for business process modeling are based on graphs consisting of a number of nodes and edges. Both nodes and edges often may be of several types. The nodes represent, e.g., different activities and events. The relations between the nodes are modeled by edges. Thus, the flow of control can be represented as a net of nodes connected by edges. The usage of the model elements is sometimes supplemented by modeling recommendations. However, all process models created by such a net-based approach are at least potential "spaghetti-processes", i.e. they are likely to have an unstructured flow of control causing deadlocks and anomalies. At the time the process is modeled first its syntactic and semantic correctness may be achieved by the discipline of its modeler. Later, when more modelers (or information systems) simultaneously and independently want to adapt and extend the model or if they want to reuse some of its parts, it is likely that they can not do this efficiently and effectively because of the deficiencies of the applied meta model.

A similar problem existed in the field of programming languages and was solved by introducing the concept of structured programming [4]. The idea of nested blocks, which always have only one entry and one exit point and therefore always lead to flows considered as being linear, can also be applied to business process modeling. Actually, each process has only one starting and ending point. These points do not refer to the several triggers of the process, but they refer to the fact that the process has started or finished. This point of view is not new: network planning techniques have used it since a long period of time.

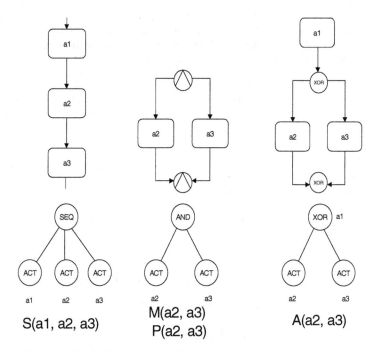

Figure 1. Blocks for process modeling as net fragment, tree and term

According to the concept of linear programming we can use blocks in order to model the process flow and the activities embedded in this flow. In contrast to nets, which have an implicit flow of control, the flow of control is explicit and a distinction between control flow (operators) and embedded activities (operands) can be made. Only a few different types of blocks are necessary to represent an suitable set of linear process models. The field of graph grammar provides syntax definition formalisms for graphical languages. Thence dividing nets into net fragments consisting of a number of nodes and edges is known. A logic block can be considered as such a fragment. Figure 1 depicts three blocks considered as basic building blocks of linear process modeling. Each block stands for a certain kind of control flow, which is why they are called logic blocks, and contains other blocks again. In the following we describe these basic logic blocks.

The sequence composition (also called production) stands for executing activities one after another in a defined order. In diagram languages sequence composition is modelled by connecting the nodes of a sequence by arrows. The approach of linear process modelling represents a sequence composition as a logic block determining the exact order of blocks contained within it. This kind of block is called sequence block and is the operator *S* in the process algebra below.

The parallel composition - concurrency and merge - means a flow of control in which all embedded activities have to be performed. In contrast to sequence composition the order is undetermined in case of a merge or, in case of concurrency, it is intended to perform the activities in parallel. In diagram languages both cases are modeled by a split node and a corresponding join node. If the join node is omitted no convergence of the branched process flows takes place; therefore planning and execution of further activities could be affected unintentionally, e.g., by deadlocks and anomalies. Additionally, more than one exit point may result.

sorts:	Activities, Processes Activities $\subset$ Processes
consts:	Activities
ops:	A, S, P, M: P x ... x P $\rightarrow$ P
eqns:	$A(x, y) = A(y, x)$ $A(A(x, y), z) = A(x, A(y, z)) =: A(x, y, z)$ $A(x, x) = x$ $S(S(x, y), z) = S(x, S(y, z)) =: S(x, y, z)$ $S(A(x, y), z) = A(S(x, z), S(y, z))$ $P(x, y) = P(y, x)$ $P(P(x, y), z) = P(x, P(y, z)) =: P(x, y, z)$ $P(x, A(y, z)) = A(P(x, y), P(x, z))$ $M(x, y) = A(S(x, y), S(y, x), P(x, y))$

Table 1. Basic Algebra

Branch node and corresponding merge node together build a logic block for parallel composition. In dependence on the intended semantics the approach distinguishes two kinds of blocks: merge blocks and concurrency blocks. The merge block means interleaving: n blocks can be performed in $n!$ different orders. If communication between the embedded blocks or the very same preconditions and starting point are needed, the concurrency block is used. In the algebra below, merge is the operator M and concurrency is the operator P. Like the parallel composition the alternative composition (also called sum) consists of several control flows between a split node and a corresponding join node. In contrast to parallel composition, a selection of blocks inside the block takes place. The basic form selects only one block out of the set of embedded blocks. This kind of logic block is called alternative block. It is the operator A in the process algebra used here.

Table 1 shows an algebra whose operators are the introduced logic blocks. This basic algebra is related to the process algebra [2], excepting it uses n-ary prefix operators instead of binary infix operators. Besides the listed operators some further operators seem to be missing, e.g., there are net fragments having semantics like a logical OR or repeat embedded blocks. These logic blocks can be assembled from the basic ones. If a logic block that is able to select one up to all embedded blocks is wanted a block O with a combination of parallel and alternative composition is used. A loop L can be considered as a special kind of alternative block that has one reverse path inside. The corresponding operators can build up from basic operators (e represents the empty block):

Merged alternative block: $O(x, y) = A(x, y, M(x, y))$
Loop block: $L(x) = A(S(a, L(x)), e)$

The basic algebra does not include left distributivity of S over A, because this leads to a difference in branching structure [2]. Without full distributivity each process model has a head normal form. If the difference in the moment of choice caused by full distributivity is out of interest each process has a disjunctive normal form [8]. This form can be very useful as a schema for storing or comparing cases.

Applying Generic Blocks

All the blocks introduced above have only one entry and one exit point regardless of whether they represent one kind of control flow or an activity or a whole process model. According to this all these different kinds of blocks can be generalized into a kind of block that is called generic block [1]. This allows us to use generic blocks as operands of logic blocks. When modeling complex control flows, generic blocks are instantiated by other blocks, thus achieving a nested structure.

In a grammar describing this way of process modeling each process model starts from a single generic block. This block marks the entry and exit point of the process. By substituting generic blocks with logic blocks we get a structure of nested blocks that represents the control flow of a business process. Finally a generic block may be substituted by an activity. In a grammar the activity would represent the terminal symbol. A block that does not contain a generic block will be called concrete block otherwise generic block. Analogously, a process model that does not hold a generic block will be called concrete process model.

Genericity permits to define parameterized blocks respectively process models. Such generic blocks are not applicable directly, but they are intended to define templates that are further used to define concrete blocks. The embedded generic blocks are parameters of its enclosing generic block; they form the template signature. By assigning concrete blocks to all of these parameters a concrete block is instantiated from the template. In case that generic blocks are assigned or parameters are left out a further and more complex generic block is defined. The instantiation of generic blocks can be controlled using constraints [1]. Besides the flexibility by selection reached by alternative blocks we also achieve flexibility by adaptation, controllable for all process models regardless of their scope and complexity.

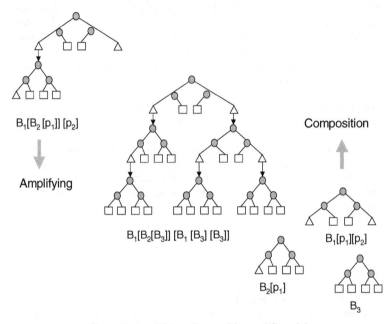

Figure 2. Amplifying, Composition and Genericity

The complete amplifying of a block by instantiating generic blocks with activities and all parameters with concrete blocks ensures that computer based execution of the process holding this block can be performed. The most detailed modeling level for the process is reached. Between this level and the level of considering the process in the aggregate a lot of amplifying levels can be defined. Such levels are interesting for analyses, because for some parts of a process an aggregated view is sufficient while for other parts a detailed view is needed.

Visualization and Exchange

An advantage of diagram languages is their vividness. This is an advantage of generic linear process modeling, too. Here a block has three forms of representation: diagram, tree structure and term. A diagram consists of the net fragments standing for blocks. In tree form each block is represented by a node connected by edges with its embedded blocks. The root node represents the logic block containing the whole structure. In the case of concrete process schemata, all leaf nodes are activities; otherwise there is at least one leaf node being a generic block. Logic blocks represent the process flow, between root and leaf nodes. To interpret a tree, it must simply be traversed. The tree form is similar to hierarchical network plans [7]. The term form is useful for algebraic purposes or if a text string is needed. All three forms of representation are semantically equivalent.

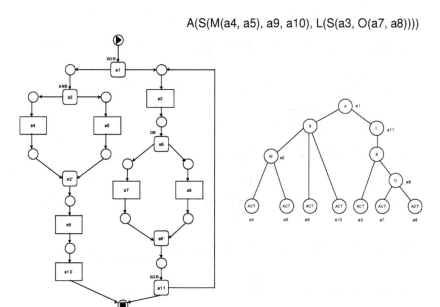

Figure 3. Concrete process schema as conventional diagram, as process algebra term and as tree

Today, in electronic commerce XML is used for data exchange in general. It is also used for exchanging business process models respectively their definitions. The Workflow Management Coalition has already issued a Workflow Process Definition Language [9]. It is intended to exchange process models based on different net approaches or diagram languages between workflow management systems. Besides other elements this definition contains a list of activities and a list of transition information. Further, activities can have a transition restriction part that contains AND/XOR-join descriptions for activities with more than one incoming transitions or AND/XOR-split descriptions for activities with more than one outgoing transition. In contrast, the DTD for generic linear process models corresponds to their tree structure, which means that the different kinds of blocks are represented by different element types. Moreover, in analogy to the generic linear process models their XML documents define explicitly the flow of control, too; the same nested structure is used. Table 2 gives an insight into a DTD for generic linear process models.

```
<!-- process.dtd -->
...
<!ENTITY % BLOCK "(ACTIVITY | GENERICBLOCK | SEQUENCE | MERGE | PARALLEL
| MERGEDALTERNATIVE |          ALTERNATIVE | LOOP)">

<!ELEMENT PROCESS (%BLOCK;)>
<!ATTLIST PROCESS     TITLE CDATA #REQUIRED ... >

<!ELEMENT ACTIVITY EMPTY >
<!ATTLIST ACTIVITY     TITLE CDATA #REQUIRED ... >

<!ELEMENT GENERICBLOCK EMPTY>
<!ATTLIST GENERICBLOCK TITLE CDATA #REQUIRED ... >

<!ELEMENT SEQUENCE ((%BLOCK;), (%BLOCK;)+)>

<!ELEMENT MERGE((%BLOCK;), (%BLOCK;)+)>

<!ELEMENT PARALLEL((%BLOCK;), (%BLOCK;)+)>

<!ELEMENT MERGEDALTERNATIVE ((%BLOCK;), (%BLOCK;)+)>
<!ATTLIST MERGEDALTERNATIVE          RULE CDATA #REQUIRED ... >

<!ELEMENT ALTERNATIVE ((%BLOCK;), (%BLOCK;)+)>
<!ATTLIST ALTERNATIVE          RULE CDATA #REQUIRED ... >

<!ELEMENT LOOP (%BLOCK)>
<!ATTLIST LOOP          RULE CDATA #REQUIRE ... >
...
```

Table 2: DTD fragment

Concluding Remarks

The presented approach does not claim to be complete. The main objective is to give a modeling approach enhancing adaptability, support of composing and amplifying, ability of automatic execution/interpretation and exchangeability of process models. Additionally, the vividness of diagram languages should to be preserved. Certainly, there are a lot of further aspects of business process models not treated here. E.g., the set of block types respectively operators is surely incomplete for practical purposes. Additional block types with different semantics will be found and appended. Ultimately, like an entity a business process is an business object too [3]. Therefore the approach is expected to develop into an object oriented approach of business process modeling comprising process subject, owner, resources etc. The information systems used today often only represent a model of states or entities. One advantage that comes with proper business process models will be that the knowledge represented by these models can be processed, employed and stored, not only the results of applying these knowledge. In the other direction, knowledge acquisition by automatic capturing of process descriptions is an interesting application, too.

References

1. van der Aalst, W.M.P.: How to handle dynamic change and capture management information? An approach based on generic workflow models. http://wwwis.win.tue.nl/~wsinwa/genws.pf.
2. Baeten, J.C:M.; Weijland, W.P.: Process Algebra, Cambridge University Press 1990.
3. Casanove, C.: OMG Common business objects and Business Object Facility RFP. OMG Document CF/96-01-04. www.omg.org.
4. Dahl, O.J.; Dijkstra, E.W.; Hoare, C. A. R.: Structured Programming, Academic Press 1981.
5. Horn, S; Jablonski, S: An Approach to Dynamic Instance Adaption in Workflow Management Applications. http://ccs.mit.edu/klein/cscw98/paper21.
6. Jablonski, S.;Böhm, M.;Schulze, W. (Hrsg.): Workflow-Management, dpunkt-Verlag 1997. (in German)
7. König, W.; Kurbel, K.; Mertens, P.; Preßmar, D.: Distributed Information Systems in Business, Springer 1996.
8. Rittgen, P.: Prozeßtheorie der Ablaufplanung, Teubner 1998. (in German)
9. Workflow Management Coalition: Interface 1: Process Definition Interchange Process Model, Document Number WfMC TC-1016-P, Document Status - Version 1.1 (Official release), Issued on October 29, 1999; http://www.wfmc.org.

Business Modelling Is Not Process Modelling

Jaap Gordijn[1,2], Hans Akkermans[1,3], and Hans van Vliet[1]

[1] Vrije Universiteit Amsterdam
Faculty of Sciences, Vuture.net — Amsterdam Center of e-Business Research
De Boelelaan 1081a, NL-1081 HV Amsterdam, The Netherlands
{gordijn,HansAkkermans,hans}@cs.vu.nl
[2] Deloitte & Touche Bakkenist, Wisselwerking 46 NL-1112 XR Diemen, The Netherlands
[3] AKMC Knowledge Management, Klareweid 19, NL-1831 BV, Koedijk, The Netherlands

Abstract. Innovative e-business projects start with a design of the e-business model. We often encounter the view, in research as well as industry practice, that an e-business model is similar to a business process model, and so can be specified using UML activity diagrams or Petri nets. In this paper, we explain why this is a misunderstanding. The root cause is that a business model is not about process but about value exchanged between actors. Failure to make this separation of concerns leads to poor business decision-making and inadequate business requirements.

1 Introduction

An important part of an e-commerce information system development process is the design of an *e-business model*. Such a model shows the business essentials of the e-commerce business case to be developed. It can be seen as a first step in requirements engineering for e-commerce information systems.

Sometimes, an e-business model is represented using a standard process modelling method such as the UML modelling language (activity diagrams) [8,2], Petri Nets [10], IDEF₀ [1], STRIM [6] or even (in many practitioner cases) ad-hoc diagrams with some notion of activity. Such models tend to be workflow-oriented: they show the sequence of activities to be performed and sometimes the actors doing so. In addition, it can show branches in a workflow sequence, parallel threads and synchronizations. Thus, a process model shows *how* a particular business case should be carried out.

We do not at all argue that process models are not useful in an e-commerce system development project. On the contrary, a model of the interorganizational business processes is necessary to explain *how* a business model works and results in many requirements for the e-commerce information system to be developed. However, a separation of concerns is needed here. Industry projects (in the telecom, music and energy industries) and case studies we performed [3,5], show that process models are *not* a good *starting* point for identifying business stakeholder requirements. Most e-business projects start with the *design* of a business model stating *what* is offered *by* whom *to* whom, rather than *how* these offerings are selected, negotiated, contracted and fulfilled *operationally* — as is explained by a process model.

In this paper, we discuss in detail what the differences between business models and process models are. Sec. 2 discusses the various business decisions to be made in

S.W. Liddle, H.C. Mayr, B. Thalheim (Eds.): ER 2000 Workshop, LNCS 1921, pp. 40–51, 2000.

e-business design. Business models and process models refer to different decisions and requirements of different stakeholder groups. For the modelling of business processes well established methods are available, but for the description of an e-business model we find them basically lacking. Therefore, we consider in Sec. 3 what the conceptual structures are that make up an e-business model in general.

This is based on the e^3-$value^{TM}$ e-business modelling method we have developed, of which more elaborate explanations and applications are given in [3,5,4]. A key point is that an e-business model is built around the notion of *value* networks, a concept absent in process modelling. Thus, business models and process models differ in the decision support they give (i.e., the modelling goals) and in their underlying core concepts. This results in a variety of practical conceptual modelling differences, which are analyzed, illustrated with industry examples, in Sec. 4, and summarized in Sec. 5.

2   Decisions in e-Business Design

The design of e-business applications consists of at least the following processes: (1) the business model design and (2) the business process model design [3]. The design decisions represented by a business model differ from those represented by a business process model. A business model shows the essentials (the strategic intent) of the way of doing business in terms of stakeholders creating and exchanging objects of *value* with each other, while a business process model shows decisions regarding the *operationalization* of a way of doing business.

2.1   Business Modelling

Most e-business projects should start with the design of the way of doing business: the business model. Essentially, it provides the design rationale for e-commerce systems from a business point of view.

In our view, the main goal of a business model is to answer the question: "*who* is offering *what* to *whom* and expects *what* in return". Therefore, the central notion in any business model should be the concept of *value*, in order to explain the creation and addition of value in an multi-party stakeholder network, as well as the exchange of value between stakeholders. The notion of *value* as an important concept in business models is also pointed out in [9] in terms of benefits and revenues.

Consequently, the main design decisions to be represented in a business model are:

1. who are the value adding business actors involved;
2. what are the offerings of which actors to which other actors;
3. what are the elements of offerings.
4. what value-creating or adding activities are producing and consuming these offerings;
5. which value-creating or adding activities are performed by which actors.

2.2   Business Process Modelling

A business model does not state *how* value-creating activities are carried out. This is an important goal of business process modelling. Other goals of business process modelling are [6,10]: (1) creation of a common approach for work to be carried out; (2) incremental improvement of processes (e.g. efficiency); (3) support of processes by workflow management systems; (4) analysis of properties of a process (e.g. deadlock free).

To present the *how*, a business process model typically shows the following design decisions:

1. who are the actors involved in the operations;
2. which operational activities can be distinguished;
3. which activities are executed by *which* actors;
4. what are the inputs and outputs of activities;
5. what is the sequence of activities to be carried out for a specific case;
6. which activities can be carried out in parallel for a specific case.

Accordingly, the nature of design decisions to be represented in an e-business model differs from the decisions to be represented in a process model. An e-business model shows the *what* aspects: what objects of value are created for whom and by whom in multi-party stakeholder network, whereas a business process model shows the associated *how* aspects. An important general goal of conceptual modelling is to provide support for decision-making. Business models and process models thus clearly differ in the types of decisions they are able to support. The importance of separating the *how* from the *what* concerns is anathema already for a long time in conceptual modelling, and it continues to be valid in e-business modelling as ever.

3   Conceptual Structures in e-Business Modelling

3.1   A Practical Business Example

We will present a practical business example to explain what conceptual structures make up an e-business model. It is based on a real-life e-business project we carried out, and is about an e-contact service.

The Ad Association is a company that coordinates more than 150 local free ad papers called FAPs. FAPs produce traditional, 'analogue' papers with ads. They are independent, often privately owned organizations, which are located around the world. A FAP serves a geographical region, for instance a large city or a county, because most goods offered in ads only reach a regional market. However, the Ad Association expects that *contact ads* may have a broader scope, even world-wide. Therefore, the Ad Association and the FAPs have decided to exploit their already locally known brand names to set up a contact ad service with a world-wide scope. Moreover, such a service will only be available as an Internet service; contact searchers can submit an ad using their browser, and can search in the ads database via their browser.

In an e-business model we represent decisions regarding stakeholders partipating in a business (in this case the e-contact service), and the creation, exchange and consumption of value in such a multi-actor stakeholder network. Figure 1 shows a high-level business model for the contact ad business idea described above. We note that this is only one of the possible business models, and the design of and choice between several possible strategic alternatives is part of any e-business project. This model shows that contact searchers, a number of FAPs and the Ad Association are involved. More importantly, it represents decisions regarding *who* is exchanging *what* with *whom* and expects *what* in return. For instance, a contact searcher is prepared to submit an ad (and thereby giving up some privacy), and expects a desired contact in return.

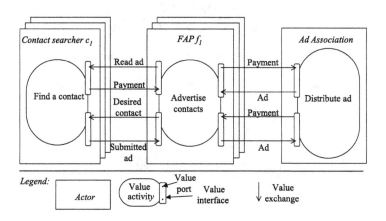

Fig. 1. A top-level business model for the Ad Association.

3.2   Generic Concepts Underlying an e-Business Model

Analysing business models for various applications such as the above contact ad one, it is possible to define a set of generic concepts and relationships that together make up an e-business model. This is depicted in Fig. 2. It forms part of our e^3-$value^{TM}$ method for e-business modelling, which is more extensively described in [3,5,4], the latter specifying an ontology for e-business models. Here, we briefly survey the core concepts, only to the extent necessary to analyse the difference between business and process models.

Actor. An actor is perceived by its environment as an *independent economic* (and often also legal) entity. By doing *value activities* (see below) actors makes profit. In a sound, viable, business model *every* actor is capable of making profit.
Value Activity. A value activity is *performed by* an actor to produce objects of value (outputs) by adding value to other objects of value (inputs). Value addition should be done against reasonable costs; if this is the case the value activity is *profitable*. Value activities *must* add value and must be profitable. The rationale for this is that during business model design, we want to study various possibilities for the assignment of value activities to actors. Actors, however, are only interested in performing activities if they are profitable for them.

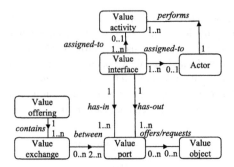

Fig. 2. Generic concepts and relationships underlying an e-business model.

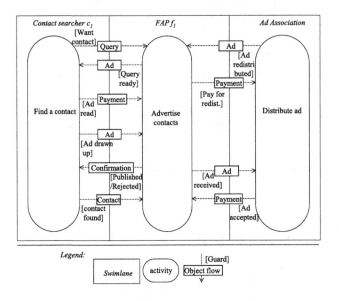

Fig. 3. A top level activity model for the Ad Association.

Value Object. Actors and value activities exchange value objects. A value object is a service, thing, or consumer experience that is of value to one or more actors. A value object is the basic building block for the creation of an offering of an actor to another actor. A value object has one or more valuation properties. These properties are used by end-consumers to determine the value they assign to an object.

Value Port. An actor or value activity uses a value port to provide or request value objects to or from its environment. Thus, a value port is used to interconnect actors in a component-based way.

Value Interface. Actors or value activities have one or more value interfaces modelling the offering of an actor or value activity to its environment. A value interface groups value ports. It shows the value objects an actor is willing to exchange in return for other value objects via its ports. The concept of value interface is based on the principle "one good turn deserves another". The value interface says nothing about the time ordering of objects to be exchanged on its ports. It simply states which value objects on which ports are available, in return for some other value objects.

Value Exchange. A value exchange represents the trade of a value object between value ports. It shows which actors are willing to exchange objects of value At least two value ports participate in a value exchange. A value port can be in multiple value exchanges. A value exchange occurs between ports of opposite direction. A value object flows from an out-port to an in-port. Therefore, at least one in-port and one out-port should be present in a value exchange.

The above constructs provide, from a conceptual modelling viewpoint, the major building blocks for designing an e-business model. Clearly, these constructs differ in significant ways from the workflow-oriented ones in process models.

4 Differences between e-Business Modelling and Business Process Modelling

E-business modelling differs in several ways from business process modelling:

1. The goal of e-business modelling is to come to a common understanding between stakeholders regarding *who* is offering and exchanging *what* with *whom* and expects *what* in return. The goal of a process model is to clarify *how* processes should be carried out, and by *whom*.
2. The concepts in e-business modelling are centred around the notion of *value*, while in process modelling concepts focus on *how* a process should be carried out.
3. The statements about the Universe of Discourse differ in e-business models and process models. An e-business model says to which extent actors add value and are profitable, and whether actors are willing to exchange objects of value with each other. A process models states which activities should performed, in which order, and which objects (in which order) should be exchanged.
4. Different model decomposition rules apply. In e-business modelling we use decomposition of value-adding activities as a way to discover new value-adding and profitable activities, for instance to discuss new alternative assignments of such activities to actors. Decomposition of activities in process modelling serves the goal of clarity, or studying various resource allocations (e.g. operational actors) to activities.

The first two differences have been discussed in the previous sections. The latter two will be elaborated in detail in this section. Here, we will take many practical examples from the contact ad case discussed in Sec. 3. For this business case example, Fig. 1 shows an e-business model, and Fig. 3 introduces an activity model. Detailed views are given in Fig. 4 and Fig. 5.

4.1   Value Object and Object

Value. In a business model, objects are only shown if they are of *value* to stakeholders. In a process model, objects are shown if they serve as required inputs of activities or are produced as outputs. As a consequence, not all objects that are part of a process model need appear in a business model, because some objects may not be of direct value to someone; and a business model may identify objects that are not present in a process model.

Example: An object present in a process model, not present in a business model. Fig. 3 shows an object *confirmation*, which models that a contact searcher receives a (positive or negative) confirmation after a submission of an ad. This *confirmation* object is not modelled in the business model (Fig. 1), because it is not of direct value to the contact searcher. It is only needed as control information, for instance to trigger the contact searcher to re-submit his/her ad after rejection.

Example: An object in a business model, not present in a process model. In Fig. 3 a *desired contact* is not present, because there is no direct corresponding physical or information object flowing from the FAP to the contact searcher. A desired contact states a consumer experience of the contact searcher; namely that s/he found a contact s/he likes. As a valuable consumer experience, it is present in the business model (Fig. 1).

Object properties. Different subsets of object properties are identified for business models and process models. A business model identifies those object properties, which can be used by a stakeholder to determine the *value* of the object, whereas object properties in process models can be used by an activity to determine a *state transition*.

Example: state transition property and value property. The *publishing date* of an ad on a website is a property useful in a process model, because it can be used to determine a state transition; from an invisible ad to a visible ad. This property, however, is not very useful in determining the *value* of an ad. Because the business model in Fig. 1 states that a reader has to pay for reading an ad, an interesting value property is the *likelihood* an ad contains a contact the reader is interested in (e.g. based on the reader's personal profile). Such a property partly determines the value a reader assigns to an ad read.

In sum, objects themselves as well as the kind of object properties differ between a business model and a process model. In a business model objects need to yield value to someone, while in a process model objects serve as inputs and outputs for activities. In a business model, properties of an object should be usable for valuing the object by an actor, whereas in an activity model properties can be used to determine a state transition.

4.2   Value Exchange and Flow

Transfer of ownership. As explained in Sec. 3, objects of value are exchanged between actors/value activities through *value exchanges*. The goal of such a construct is to model a *legal transfer* of value objects. In a process model data flows and control flows are used to model a transition from one activity to another. It is used to express *how* activities should be carried out in terms of sequences or parallelisations of activities.

Example: Flows and rights. The process model in Fig. 3 contains a data flow from the contact searcher to the FAP called *contact* that states that the contact searcher reports the experience of a desired contact to the FAP. This flow relates to the *desired contact* value exchange (Fig. 1) from the FAP to the contact searcher, but it is not the same. The *contact* flow is necessary as *control* information, for instance as a trigger to remove a published ad as soon as a desired contact occurs, while the *desired contact* value exchange models the *valuable experience* itself. Note that in the process model, the confirmation flows from contact searcher to FAP, while the desired contact flows from the FAP to the contact searcher.

No direct physical or information flow. A value exchange may coincide with a flow of a physical product or information if these are of value to a stakeholder. However, sometimes a value exchange states a consumer experience, which has no underlying direct physical or information flow. The previous example illustrates this case also.

In conclusion, a value exchange expresses a change of ownership (as an economic result, not as a process outcome), which is normally not expressed in process models. Moreover, some value exchanges do not imply a physical or information flow directly, but instead express an actors' consumer experience.

4.3   Value Interface

In a business model, we have the notion of *value interface* expressing the principle "one good turn deserves another" (a rule or law of value exchange). This allows stakeholders to clarify to each other *what* objects of value they are prepared to exchange in return for other objects; a key decision during business modelling. Such a principle is not present in process models.

Example: One good turn deserves another. From Fig. 3 it cannot easily be concluded that a reader has to pay for reading an ad, while Fig. 1 clearly shows that a *read ad* is offered in return for a *payment*.

4.4   Activity

Value adding and profitable. In process modelling, an activity denotes something to be done, in order to produce outputs as a result of inputs and resources. In a business model, we distinguish activities only if they are *profitable* for the performing stakeholder. The rationale for introducing value activities is that, for a particular business case, we want to determine the amount of profit for each actor, and we want to address and discuss various assignments of activities to stakeholders. Stakeholders are only interested in performing activities if these are profitable.

Decomposition. The different interpretations of the *activity* concept in business models and process models leads to different decompositions.

In the literature on process modelling, a number of motivations are given for the decomposition of activities into sub-activities. IDEF$_0$ [1] indicates that an activity should be recursively decomposed in 5 to 7 sub-activities, until a common understanding about the activity is reached by stakeholders. In this case, decomposition serves the goal of *clarity*. In STRIM [6], activities are decomposed until they can be regrouped and assigned to a particular role (i.e., operational actor). Decomposition then serves the goal of clarifying resources needed in carrying out tasks. In a business model, however, we only decompose a value activity if *all* resulting sub-activities themselves add value and are profitable. In [7] and [9], this is referred to as value chain deconstruction, as a way to discover new activities which can be successfully assigned to alternative commercial actors.

Example: Different decompositions.

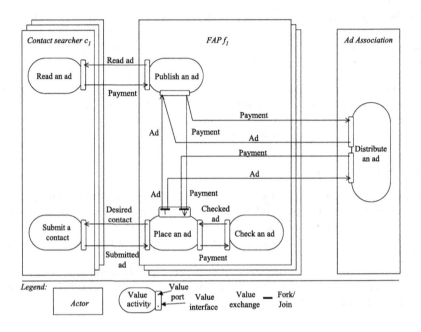

Fig. 4. A decomposed business model for the Ad Association.

Fig. 5 decomposes the activities introduced in Fig. 3. For brevity, we do not show the complete decomposition but focus on the submission of an ad. The main goal of the decomposition is to illustrate *how* a submission process should be carried out. After an ad is submitted by a contact searcher, it is checked (e.g. for absence of dirty language). If the ad passes this check, it is added to the website of a FAP and the contact searcher receives a confirmation. Also, the ad is offered to the Ad Association, which pays for it afterwards. The Ad Association supplies the ad to other FAPs. In sum, this detailled process model shows the activities necessary for a submission, as well as their execution sequence and parallel threads for an ad submission.

Figure 4 shows a decomposition of the value activities in Fig. 1 into *profitable* sub-activities. The decomposition operation is defined as follows: (1) a value activity can be decomposed in other (sub) value activities if each sub-value activity adds value and is profitable; (2) consider for each pair of sub-value activities new value interfaces and value exchanges if required.

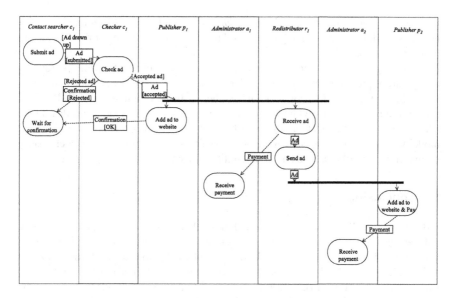

Fig. 5. A decomposed process model for the Ad Association.

The *find contact* value activity is decomposed into two sub-value activities: (1) *read an ad*, and (2) *submit an ad*. Both activities are likely to be profitable for contact searchers; they both enhance the chance to find a desired contact. The *advertise contacts* value activity is decomposed into three sub-value activities. Note that between these sub-value activities new value exchanges have been introduced. Also note that the *check an ad* activity is considered to be profitable; this is especially the case when a FAP requests another FAP to check an ad, for instance if a FAP does not speak the language in which an ad was written.

In sum, in a process model, decomposition is often led by the motivation to show a process flow in detail, while in a business model it is led by a search for commercially viable sub-activities.

4.5 Actors

Individual actors. In a process model, the actor itself is usually not shown at the instance level. At most it is indicated that a number of actors capable of performing a particular activity, should be present, for instance to model resource management. When designing business models, it should be possible to identify the profitability of a business model to a particular actor. During business modelling, these individual actors are important stakeholders. Therefore, in a business model, actors sometimes are mentioned on an individual basis.

Value adding and profitable. Actors in an process model are indicated for purposes such as resource allocation and scheduling. However, in a business model we distinguish actors to facilitate reasoning about *value addition* and *profitability*. Therefore, actors are not individual agents performing activities, but economic and legal entities that engage in business transactions.

Example: Operational actors and commercial actors. In Figure 5, actors performing acties are represented by swimlanes. The actual actor instances are not mentioned, while the business model (Fig. 1) indicates the existence of a number of FAPs which can be addressed on an individual basis. Moreover, in the business model we distinguish FAPs, being legal entities that engage in business transactions, whereas in the process model we identify resources carrying out work for such an entity, such as a checker, a publisher, a redistributor, and an administration officer.

5   Conclusions

e-Business modelling and process modelling are both forms of conceptual modelling, both are necessary for good e-business design, but they differ in several significant ways. First of all, the main goal of e-business modelling is to reach agreement amongst stakeholders regarding the question "*who* is offering *what of value* to *whom* and expects *what of value* in return". In contrast, an important goal of process modelling is to reach a common understanding about *how* activities should be carried out (e.g. in which order). These are two different modelling goals, asking for different modelling methods with different constructs. Modelling strategic intent of e-business differs from modelling operational fulfilment.

As a result, the contents of an e-business model and a process model also differ in a number of ways:

1. The concepts in e-business modelling are centred around the notion of *value*, while in process modelling concepts focus on *how* a process should be carried out in operational terms.
2. In an e-business model, an actor adds value and is profitable, while in a process model an actor performs an operational process.
3. In an e-business model, objects represent something of value to a stakeholder, while in a process model objects serve as inputs and outputs for activities and may be used the steer the process flow.
4. In an e-business model, object properties can be used by a stakeholder to determine the value of an object. In a process model, object properties are used to determine state transations.
5. In an e-business model, value exchanges represent a transfer of ownership, while in a process model a flow of information or goods implies a change of state;
6. In an e-business model, we have the notion of "One good turn deserves another", which is conceptualised by the value interface. Such a notion is absent in process modelling.
7. In an e-business model, we are only interested in activities which are capable of adding value and are profitable. Decomposition of such activities is done to discover smaller chunks of activities that still add value and are profitable. Discovering these activities often leads to re-assignment of activities to actors. In a process model, decomposition serves the goal of clarification of the workflow or to show the

assignment of activities to working actors. Hence, the model decomposition rules are different.

We have used our e^3-$value^{TM}$ method in a number of industrial e-business development projects. Our experience is that a focus on the creation and distribution of value in a stakeholder network is a convenient way to express, negotiate and clarify business models to stakeholders. Distinguishing the distribution of value (e-business modelling) from the way processes are actually performed (process modelling) leads to a separation of concerns of stakeholders and clarifies the discussions.

Acknowledgement. This work has been partly sponsored by the Stichting voor de Technische Wetenschappen (STW), project nr VWI.4949.

References

1. $IDEF_0$ method report. http://www.idef.com/Complete_Reports/idef0, 1981.
2. Martin Fowler and Kendall Scott. *UML Distilled - Applying the standard object modelling language*. Addison Wesley Longmann, Inc., Reading, Massachusetts, 1997.
3. J. Gordijn, J.M. Akkermans, and J.C. van Vliet. Value based requirements creation for electronic commerce applications. In *Proceedings of the 33rd Hawaii International Conference On System Sciences (HICSS-33)*, pages CD–ROM. IEEE, January 4-7 2000. Also available from http://www.cs.vu.nl/~gordijn.
4. J. Gordijn, J.M. Akkermans, and J.C. van Vliet. What's in an electronic business model. In *LNAI*. Springer-Verlag, 2000. To Appear. Available from http://www.cs.vu.nl/~gordijn.
5. J. Gordijn, H. de Bruin, and J.M. Akkermans. Integral design of E-Commerce systems: Aligning the business with software architecture through scenarios. In H. de Bruin, editor, *ICT-Architecture in the BeNeLux*, 1999. Also available from http://www.cs.vu.nl/~gordijn.
6. Martyn A. Ould. *Business Processes - Modelling and Analysis for Re-engineering and Improvement*. John Wiley & Sons, Chichester, England, 1995.
7. M.E. Porter and V.E. Millar. How information gives you competitive advantage. *Harvard Business Review*, pages 149–160, 1985.
8. James Rumbaugh, Ivar Jacobson, and Grady Booch. *The Unified Modelling Language Reference Manual*. Addison Wesley Longmann, Inc., Reading, Massachusetts, 1999.
9. Paul Timmers. *Electronic Commerce: Strategies and Models for Business-to-Business Trading*. John Wiley & Sons Ltd., Chichester, England, 1999.
10. K.M. van Hee. *Informations Systems Engineering - A formal approach*. Cambridge University Press, Cambridge, 1994.

Modeling Electronic Workflow Markets

Andreas Geppert

Department of Information Technology, University of Zurich
geppert@ifi.unizh.ch

Abstract. Market-based inter-organizational workflow management integrates workflow management and business-to-business (B2B) e-commerce; it is a novel approach to workflow management which regards workflows as goods traded on an electronic market. In this paper, we investigate modeling aspects of electronic workflow markets. The workflow marketplace is modeled as a federation of workflow systems. Providers export workflows to the federation, and clients import workflows offered by other parties. Providers model prices and execution time of workflows, and requestors model budgets and deadlines for imported workflows. This information is then exploited at runtime for trading workflows on the market. In particular, cost and time requirements are used to determine the optimal provider based on the current execution state of the requestor.

1 Introduction and Motivation

Workflow management [6] has recently found great attention in the information systems field, as it allows workflows to be defined in a formal language/framework and to be enacted according to their specification. A recent trend is to build *virtual enterprises*, e.g., by modeling and executing inter-organizational workflows. Inter-organizational workflow management, however, raises numerous open research problems, such as how to model these workflows and the virtual enterprise in which they are executing, how to provide for the adequate infrastructure to resolve heterogeneity such as different used workflow management systems.

Inter-organizational workflow management yields many opportunities for enterprises to redesign and optimize their processes and the corresponding workflow specifications. Using the workflows offered by others, an enterprise does not have to implement such a workflow itself—instead, it can import the workflow execution from another enterprise. Often, many providers with different workflow implementations of the same service will compete, so that the question arises how to select the best one. Thus, the real-world business is characterized by the presence of multiple competing providers, and the assessment of offers as well as the selection of cheapest/best providers has to be supported.

Hence, inter-organizational workflow management requires an economic perspective that allows to quantify costs of workflow execution (such that decisions like "make or buy" are facilitated). The selection of optimal providers in case of multiple eligible ones must also be possible based on economic criteria (such as cost, time, and quality). These objectives are accomplished by enhancing a workflow system with electronic market mechanisms, where workflows offered by the parties are the traded goods [8, 9].

In this paper, we focus on the modeling aspects of such a workflow market. A federation of workflow systems is achieved by introducing a workflow market-broker,

S.W. Liddle, H.C. Mayr, B. Thalheim (Eds.): ER 2000 Workshop, LNCS 1921, pp. 52–63, 2000.
© Springer-Verlag Berlin Heidelberg 2000

which maintains a repository containing information about workflow types offered on the market. Federation parties can buy workflows from the market, whereby the best provider is determined by a bidding protocol. Finally, we propose to extend workflow models by attributes (of steps) such as execution cost and time; these properties are then used to optimize workflow execution with respect to deadlines and budgets.

The remainder of this paper is organized as follows: in the next section, we introduce necessary terminology and give an overview about our approach towards market-based inter-organizational workflow management. In section 3, we discuss modeling aspects of workflow markets, and section 4 briefly presents market-based workflow execution. Section 5 surveys related work, and section 6 concludes the paper.

2   Market-based Workflow Management: Overview

In this section, we first introduce the relevant terminology, and then give an overview of our approach. A *workflow type* is an executable representation of a (business) process. The structure of the workflow type defines the steps (atomic activities or subworkflows) necessary to enact the process. Steps can be connected by control flow constructs (e.g., sequence) and data flow constructs. Activities are executed by *processing entities* (PEs), which can be humans (e.g., clerks) or software systems. *Task assignment rules* define which one of the eligible PEs actually is requested to execute an activity. A workflow type defines an *interface*, consisting of a unique name and sets of typed formal input and output parameters, and a *body* which implements the workflow, i.e., defines subworkflows/activities, task assignment rules, involved PEs, etc.

A *workflow management system* (WFMS) supports the definition and execution of workflow types. A *workflow system* (WS) consists of a WFMS and a set of workflow types. A *federated WS* (or *federation*, for short) consists of a set of WS. The underlying individual WFMS may be heterogeneous with respect to the workflow models they use, underlying data stores, etc. Since the parties involved in the execution of workflow can belong to different enterprises, we consider the federation as the workflow system of a *virtual enterprise*.

We present a running example using the TRAMs workflow specification notation [10] in Fig. 1. The company Innova SA specifies a workflow for evaluating new product concepts. Initially, the marketing department develops a new product concept (activity newConcept). Then, a concept evaluation must be performed (conceptTest), and a marketing plan is developed in which price, distribution, and positioning of the new product are determined (marketPlanning). After successful concept testing and completion of the marketing plan, an extensive market survey is performed (marketSurvey). As soon as the survey is completed, its results are evaluated and following the profit/cost estimation (profit/costEstimate) the go-ahead decision is taken (decide). Note that the concept test and market survey are themselves complex processes, which are outsourced according to current business practice.

Market-based workflow management treats workflows and/or activities as goods traded on an electronic market. It has been proposed originally for trading atomic activities in the intra-enterprise context [8]. Its application to inter-organizational workflow management has been outlined in [9]. The principle of market-based workflow manage-

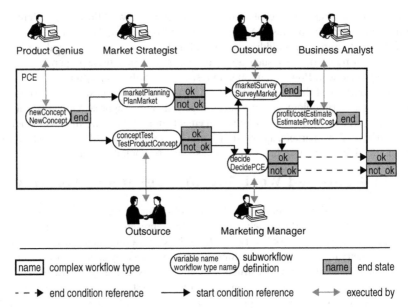

Figure 1 The Product Concept Evaluation (PCE) workflow type.

ment is to establish a market in which steps of a workflow are traded goods. Thus, WS act as consumers by buying services from other WS (the sellers). Each service offered by a seller is characterized by the time period the WS needs to execute it, by a price the seller charges, and by a domain-specific notion of quality the seller guarantees.

The market mechanism helps to optimize each single workflow execution in such a way that total execution time and overall aggregated costs are balanced and remain within predefined limits, while simultaneously the requirements to QoS of each workflow step are met. This objective is achieved by using a bidding protocol for task assignment, and by taking the required quality levels as well as expenditures and execution time of the workflow into account when selecting the best bidder for a concrete step to be executed.

3   Modeling Workflow Markets

In this chapter the specification of information necessary to enable the trading of workflows is addressed. A workflow `wf` is called *inter-organizational* if at least one of its subworkflows is executed by a WS different from the one `wf` is enacted by. To enable the specification of inter-organizational workflows by a WS, the following three prerequisites have to be met:

1. it must be possible for a WS to find out which workflow types are offered,
2. workflow types offered by a certain WS can be represented in another WS,
3. software components must be representable as PE, and task assignment rules must be specifiable (note that this functionality is fairly standard).

3.1   Repository-based Modeling of Workflows

Workflow systems can use (and, as we will show below, also buy) workflows offered by other WS on the market. This market is managed by a *workflow market broker* (WMB), which maintains all the information necessary to enable the trading of workflows in such a virtual enterprise. In this section, we discuss how to model the functionality offered and the data managed by the WMB.

The WMB maintains a repository of all workflow types offered in the federation by one or more participating WS. This broker offers trading functionality [e.g., 15, chapter 16], i.e., it allows workflow types to be announced by exporters. Clients (i.e., importers) can query the broker in order to retrieve information about workflow types offered in the federation. In addition to the trading functions, the broker also supports proper brokering functionality (see below). The broker maintains a repository, in which it stores information about the workflow types exported by federation components. This repository contains the following information for each exported workflow type:

- its name,
- a set of typed formal input parameters,
- a set of typed formal output parameters,
- a set of WS that implement and export this workflow type,
- a set of WS that import the workflow type,
- a documentation, which explains what a workflow type does, without providing implementation details,
- further market-related information (see below).

The WMB offers a set of services to the workflow federation with which potential clients can "buy" offered workflows. The provider bidding and subsequent selection is performed by the WMB. These services are grouped into three kinds of interfaces, a buildtime interface (interface 1 in Fig. 2), the runtime customer/WMB interface (interfaces 2 and 5), and the runtime provider/WMB interface (interfaces 3 and 4). The offered services

- provide an export interface to the providers and buildtime information to workflow federation parties about the actual workflow market (interface 1);
- accept a request from a client for a workflow with specific price, duration, and quality requirements (interface 2); this initiates a bidding process (interface 3);
- accept and evaluate bids from potential workflow providers (interface 4);
- accept workflow execution replies (interface 4) from providers in order to forward them to customers (interface 5).

The interfaces relevant for runtime are discussed and illustrated below. The services relevant for buildtime support:

1. querying the repository,
2. inserting new workflow types into the repository,
3. exporting workflow types, and
4. importing workflow types.

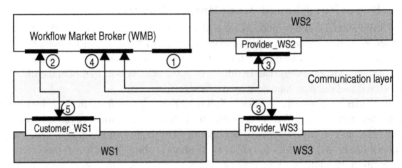

Figure 2 Architecture of a market-based workflow federation

Workflow systems can query the repository to find out about already defined workflow types. These queries are crucial in two respects:

- federation parties can determine whether one of the workflows they are willing to offer already exists in the federation,
- federation parties that want to import a workflow type from another party can check whether the workflow type is offered on the market at all.

In case a workflow type does not yet exist, a federation party can ask the WMB to insert it into the repository. To that end, it has to provide the workflow name and interface as well as the documentation describing the semantics of the workflow type.

Finally, workflow systems can export a workflow type to the federation. When exporting a workflow type WT, the workflow system has to send the following information to the WMB:

- its (i.e., the exporter's) identity,
- the name of the workflow type.

Exporting a workflow type establishes a contract in that the workflow system states that its workflow type definition complies to the interface as defined in the repository, and that the (local) implementation of the workflow type obeys the semantics as represented in the documentation given in the repository.

Note that in our approach only workflow type interfaces are exported and imported. Thus, the body of a workflow is not visible to importers (this is in contrast to other proposals [1, 3, 17, 20]). In practice, workflow specifications define how an enterprise conducts its business, and an optimized, efficient workflow specification will represent a competitive advantage. Hence, workflow specification internals represent a strategic asset that enterprises would not like to exhibit to other companies, especially not to competitors.

In our example, we assume that the step named `marketSurvey` of type `SurveyMarket` is outsourced. Two market research companies, Exel Market Co. and Good Surveys Ltd., export a workflow type named `TestProductConcept`. They also compete in performing market surveys by exporting the workflow type `SurveyMarket`. Table 1 shows the information about the workflow types `TestProductConcept` and `SurveyMarket` that is contained in the repository.

Table 1: Repository Information about exported workflow types

WFType	In-Parameters	Out-Parameters	Exporter/ Provider	Importer /User	Quality Criteria
Test-Product-Concept	customer: string, concept: Document	testResult: Document	Exel Market Co. Good Surveys Ltd.	Innova SA	testType: {sketch, mock-up}
Survey-Market	customer: string, concept: Document	surveyResult: Document		Innova SA	sample: {random, non-random} interview: {personal, mail, phone}

3.2   Modeling Providers and Clients

The execution of workflows offered by other federation parties requires appropriate modeling constructs as well as infrastructure extensions, because workflows purchased on the workflow market might (and usually will) run at a different location, on another operation system and, last not least, using a different WFMS. On the client side, it must be possible to refer to workflow providers in such a way that access to the market and outsourcing of workflows is possible, while the client should not have to bother for details such as the providers' WFMS, etc. This is accomplished by introducing a generic type of processing entity which acts as a mediator between the client workflow system and the market broker. From the viewpoint of the local WS, this PE cannot be distinguished from other (fully local) PEs.

Furthermore, a workflow system must be able to include workflow types imported from the repository such that they can be instantiated by the workflow system as any other locally defined workflow type. Such imported workflow types are modeled as automated, atomic activities. "Atomic" means that the activity has no internal structure, or at least its structure is not known to the WFMS. "Automated" means that the activity can be executed by a software system without interaction with humans.

Any such imported workflow can, therefore, be modeled like an already existing, local activity or workflow. It can be connected to other activities or sibling workflows in terms of control flow, data flow, and so forth. The fact that it is imported is modeled by defining the aforementioned PE as being capable of executing this workflow. Whenever the workflow is to be executed, role resolution and task assignment will be done, and that PE will be identified as the one which should execute the workflow.

On the server side (i.e., the remote WS where subworkflows of a global workflow can be executed), specific clients mediate between the rest of the federation and the local WFMS. These clients are able to instantiate workflows their WS has exported, to collect results when workflows terminate, and to send these results back to the WMB. Their second major function is to participate in bidding, i.e., to accept requests for bids, compute bids, and return the bids to the WMB (see below). They need to be implemented once per WFMS and can then be reused for any exported workflow type.

The two types of mediating components are plugged into each of the component WS that should participate in the federation and use a common model of messages and parameters. The introduction of customers and providers means that the former do not need to know the concrete interface of the WFMS offering workflows—they simply have to know the service interface of the providers. Similarly, providers to not need to know interfaces/formats of their clients' WFMS, but only need to use the API of their local WFMS. In this way, customer and provider components have to be written once for each type of WFMS.

The concrete choice how to represent and implement customers and providers and the concrete communication software is of minor concern—provided that they all run on all platforms involved in the federation. For instance, an ORB [16] may be used for communication; in this case it is quite natural to use the OMG object model for representation and transmission of parameters. The component WFMS is then represented as an application object, providers implement skeletons, and customers implement stubs (in CORBA terminology).

3.3   Specifying Information Relevant for Trading Workflows

Assume a WS outsources a workflow, and several providers offer it, however at varying costs and execution times. Thus, which provider should be chosen? The natural answer is to choose the cheapest provider if the budget so far has been overdrawn, but deadlines have been met (i.e., staying within the budget is the current problem). Alternatively, if deadlines so far have been missed, then it might be favorable to spend more money in order to save time.

To enable such decisions, a workflow buyer needs information about the workflow state with respect to expenses and execution time. In particular, it must be possible to compare planned costs and execution times with the actual ones. We therefore propose to enhance activity specifications in workflow specification with two new attributes: expected execution time and planned execution cost. Several options exist when and how in the course of a business process life cycle [6] to determine these parameters:

1. during process modeling,
2. during process/workflow simulation,
3. as a result of workflow monitoring and analysis,
4. by querying the WMB.

In the first variant, results of activity based accounting are used to determine the cost incurred by the execution of each activity. This is, the cost of activities is obtained as a result of process modeling.

In the second option, execution cost of activities and workflows are derived during process simulation. These two options are already offered by some process modeling tools, such as Holosofx [14] and MetaSoftware's Workflow Analyzer [12]. Thus, in this respect no extensions are necessary on the conceptual (process-modeling) level.

The third option is to consider previous executions of the workflow type and to determine execution cost and time by analyzing the workflow execution log. At least the execution time can be easily derived from this log. Note further that it is reasonable to assume at least for providers that a significant number of workflow executions is avail-

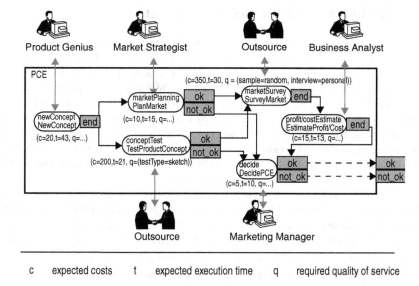

Figure 3 The Product Concept Evaluation (PCE) workflow type

able for analysis. This is, we assume that a provider has already executed a workflow type numerous times before it is offered on the workflow market.

Finally, market prices as well as execution costs can also be obtained from the WMB.

The information gained in this step is used in two ways:

- it provides sellers with the information required for specifying bids, and
- it enables clients to specify deadlines and budgets.

On the client side, a workflow specification is enhanced in that for each step, expected execution time and cost are added. During workflow execution, this information is used for assessing the current workflow executions state (i.e., whether the workflow stays within time and cost limits, or whether one/both of them is overdrawn).

Fig. 3 represents the example Product Concept Evaluation (PCE) workflow type defined within the WS of Innova SA using the TRAMs workflow specification approach [10]. As a first step in the creation of PCE, the types of the subworkflows conceptTest and MarketSurvey to be outsourced have to be imported from the repository. Next, estimates of the execution cost, execution time, and required quality of service for the subworkflows are specified. In this case, the estimates for subworkflows to be outsourced (marketSurvey and conceptTest) are derived by querying the market broker. The estimates for the rest of the subworkflows are defined based on internal information available by Innova SA, such as prior executions, simulations, and activity based costing.

4   Market-Based Workflow Execution

In this section, we briefly address market-based workflow execution in federated workflow systems. Provided that workflows have been specified and federations modeled as described above, market-based workflow execution is supported by the bidding protocol. The bidding protocol proceeds as follows: the client WS posts a request to the service market by communicating its requirements to the market broker through the customer component. The market broker reacts by selecting potential supplier WS from the federation and communicating them the posted workflow requirements through their provider components. The bids are subsequently collected by the WMB who then chooses the winning bidder and notifies all bidders of the selection result. The winner then starts the execution of the requested workflow.

In order to find the optimal bid with respect to the current workflow execution state, information about the current (aggregated) execution cost and duration of the requesting workflow instance have to be taken into account, and thus have to be computed at runtime. The following measures are required for each executing workflow instance wf:

- $C_{real}(wf, swf)$ are the actual costs spent in wf until swf's scheduling;
- $C_{planned}(wf, swf)$ are the costs planned for wf until swf's scheduling;
- $T_{real}(wf, swf)$ is the actual execution time of wf until swf is scheduled;
- $T_{planned}(wf, swf)$ is the expected execution time of wf until swf is scheduled.

The detailed description how these measures are computed, is beyond the scope of this paper (see [8]). Here it should be sufficient to state that the WFMS engine consecutively computes them after the completion of each activity.

Example. Assume a PCE-workflow `wf` is executing. The `newConcept` activity has been executed within 40 time units at a price of 25 currency units. `MarketPlanning` terminates after 22 time units (at costs of 12 currency units), followed by `conceptTest` (after 32 time units, at costs of 180). For the next step, `marketSurvey`, we obtain the cost and time measures as shown in Table 2.

Table 2: Aggregated cost and time for a sample workflow execution

Step	$T_{Planned}$	T_{Real}	$C_{Planned}$	C_{Real}	TOR	COR
newConcept	-	-	-	-	1.0	1.0
marketPlanning	43	40	20	25	0.93	1.25
conceptTest	43	40	20	25	0.93	1.25
marketSurvey	64	72	230	217	1.11	0.94

The four measures are used to determine whether and to which extent the workflow execution has exceeded time and/or cost limits, i.e., they are used to compute the amount of time or money saved or overdrawn. Based on these computations, further assignments can then stress cost or time.

The *cost overrun ratio* and the *time overrun ratio* of wf_A are defined as

$$COR(wf, A) = \frac{C_{real}(wf, A)}{C_{planned}(wf, A)} \qquad TOR(wf, A) = \frac{T_{real}(wf, A)}{T_{planned}(wf, A)}$$

For activities which have no predecessors, COR and TOR have the default value 1. In our example, we obtain the overrun rations shown in Table 2. In order to determine the best bid, these have to be ranked in such a way that cost and time are weighted according to the overrun factors. Thus, one possibility is to evaluate the n bids (C_i,T_i) according to the following formula:

$$winning\_bid = min\left(C_i^{COR(wf, SWF)} \times T_i^{TOR(wf, SWF)}\right)$$

COR and TOR are used in the exponents in order to weight cost and/or time.

In case the WMB cannot determine any eligible WS (because none can provide for the required quality, or can stay below specified time and cost limits), then it informs the requestor that (and why) no provider could be found.

In our example, we assume that two bids were collected by the WMB for the `marketSurvey` subworkflow satisfying the minimal quality requirements:

Bid 1: (c = 400, t=25, q = (sample = random, interview = personal))
Bid 2: (c = 330, t=30, q = (sample = random, interview = personal))

The winning bid is calculated as min(Bid 1: $400^{0.94}$ x $25^{1.11}$ = 9'946, Bid 2: $330^{0.94}$ x $30^{1.11}$ = 10'162) which is Bid 1. Note that for each single bidding process, it is determined whether either cost or time are more critical for the workflow in question. The fact that for a workflow cost or time are weighted more can change arbitrarily often within its execution. Table 3 contains the expected COR and TOR values before `profit/costEstimate` depending on the bid chosen. The bid selection formula aims at minimizing the sum of the deviations of TOR and COR from the ideal value of 1.

Table 3: Overrun ratios values for a sample workflow execution

Step	TOR(Bid 1)	COR(Bid 1)	TOR(Bid 2)	COR(Bid 2)
profit/costEstimate	1.03	1.06	1.09	0.94

5   Related Work

In this section, we survey related work in the areas of market-based inter-organizational workflow management. Both areas are still maturing, and we are not aware of any other work that combines both, market mechanisms and inter-organizational workflow management in the sense of this paper.

Work done by the Workflow Management Coalition (WfMC) [20], by the Object Management Group [17], and on the Simple Workflow Access Protocol (SWAP) [19] tries to establish standards in the area of workflow system interoperability. In general, these efforts are concerned with interoperability on a lower level than the work reported in this paper. Furthermore, none of these efforts considers market-based workflow management.

The CORBAservices [15] specify a trader service similar to the workflow market broker proposed here. However, in contrast to our proposal, the trader service does not cover bidding.

Inter-enterprise workflow management and workflow interoperability have also been investigated in several research projects [1, 2, 3, 7, 11, 18]. However, none of these efforts considers workflow markets.

Finally, it should be noted that, orthogonal to our proposal of applying e-commerce principles to workflow management, some projects investigated workflow management support for e-commerce [4, 13].

6 Conclusion

This paper introduced a novel approach towards market-based inter-organizational workflow management. The workflow market is formed by a federation of WS, in which the parties demand and offer the execution of workflows. Workflow trading is implemented by a market broker, which in its repository maintains information about offered workflow types. We have shown how to model exported and imported work-flow types. We also have proposed to extend workflow specifications by information about deadlines, budgets, and quality requirements. We have discussed workflow execution, which based on a bidding protocol facilitates workflow enactment in a market-based way.

Thus, the contribution of this work is that it allows to consider workflows from a market perspective, and that it leverages the concept of electronic markets to inter-organizational workflow management. We are convinced that the integration of market mechanisms and inter-organizational workflow management is of paramount importance for computerized B2B operations, because otherwise the economic perspective of involved enterprises cannot be adequately reflected on the workflow management level.

For the integration of (component) WS into the federation, we have so far assumed that the exporting parties agree on workflow type interfaces, and that these interfaces together with documentations and further definitions (such as quality criteria) provide enough information for clients to find out whether a specific workflow type is useful for them. These two assumptions will be loosened in our future work. For the integration/ export step, we will investigate how semantic heterogeneity of workflow types can be resolved (e.g., heterogeneous interfaces of workflow types with the same purpose). With respect to the degree of information about workflow types maintained by the repository, we will investigate how workflow types can be described in such a way that clients can unambiguously decide whether the workflow type is what they are looking for, while information confidential and valuable for the provider is not exposed.

References

[1]   W.M.P. van der Aalst. Process-oriented Architectures for Electronic Commerce and Inter-organizational Workflow. Information Systems, 2000.

[2]   G. Alonso, U. Fiedler, C. Hagen, A. Lazcano, H. Schuldt, N. Weiler. WISE: Business to Business E-Commerce. *Proc. 9th Int'l Workshop on Research Issues on Data Engineering*, Sidney, Australia, March 1999.

[3]   F. Casati, S. Ceri, B. Pernici, G. Pozzi. Semantic WorkFlow Interoperability. *Proc. 5th Int'l EDBT Conf.*, Avignon, France, March 1996.

[4]   A. Dogac, I. Durusoy, S. Arpinar, N. Tatbul, P. Koskal, I. Cingil, N. Dimililer. A Workflow-based Electronic Marketplace on the Web. ACM SIGMOD Record 27:4, December 1998.

[5]   A. Dogac, M.T. Oezsu, O. Ulusoy (eds). *Current Trends in Database Management Technology*. Idea Group Publishing, Hershey, PA, 1999.

[6]   D. Georgakopoulos, A. Tsalgatidou. Technology and Tools for Comprehensive Business Process Lifecycle Management. In A. Dogac, L. Kalinichenko, M.T. Oezsu, A. Sheth (eds). *Workflow Management Systems and Interoperability*. Springer, 1998.

[7]   D. Georgakopoulos, H. Schuster, A. Cichocki, D. Baker. Managing Process and Service Fusion in Virtual Enterprises. *Information Systems* 24:6, 1999.

[8]   A. Geppert, M. Kradolfer, D. Tombros. Market-based Workflow Management. *Proc. Int'l IFIP Working Conf. on Trends in Distributed Systems for Electronic Commerce*, Hamburg, Germany, June 1998.

[9]   A. Geppert, M. Kradolfer, D. Tombros. Trading Workflows on Electronic Markets. In [5].

[10]   M. Kradolfer, A. Geppert, K.R. Dittrich. Workflow Specification in TRAMs. *Proc. 18th Int'l ER Conf.*, Paris, France, November 1999.

[11]   H. Ludwig, Y. Hoffner. Contract-based Cross-Organisational Workflows - The CrossFlow Project. *Proc. Workshop on Cross-Organisational Workflow Management and Coordination*. San Francisco, CA, February 1999.

[12]   http://www.metasoftware.com/.

[13]   P. Muth, J. Weissenfels, G. Weikum. What Workflow Technology can do for Electronic Commerce. In [5].

[14]   J.E. Olson. Best Practices in Electronic Commerce Integration. White Paper, Holosofx, El Segundo, CA, June 1999 (www.holosofx.com).

[15]   CORBAservices: Common Object Services Specification. OMG, July 1997 (http://www.omg.org/corba/sectran1.htm).

[16]   The Common Object Request Broker: Architecture and Specification. Revision 2.1, Object Management Group, August 1997.

[17]   Workflow Management Facility. OMG Business Object Domain Task Force, BODTF-RFP 2 Submission, August 1997.

[18]   A. Sheth, K. Kochut, J. Miller, D. Worah, S. Das, C. Lin, D. Palaniswami, J. Lynch, I. Shevchenko. Supporting State-wide Immunization Tracking using Multi-Paradigm Workflow Technology. *Proc. 22nd Int'l Conf. on Very Large Data Bases*, Bombay, India, September 1996.

[19]   Simple Workflow Access Protocol, http://www.ics.uci.edu/~ietfswap/, 1999.

[20]   Interface 4 - Interoperability - Abstract Specification. WFMC-TC-1012, Workflow Management Coalition, October 1996

Building Multi-device, Content-Centric Applications Using WebML and the W3I3 Tool Suite

Angela Bonifati[1], Stefano Ceri[1], Piero Fraternali[1], and Andrea Maurino[1]

Politecnico di Milano,Piazza L. da Vinci 32
20133 Milano, Italy
{bonifati,ceri,fraterna,maurino}@elet.polimi.it

Abstract. In the forthcoming years, two factors will jeopardize the deployment of Web applications: supporting multi-device outputs and one-to-one personalization. These two factors will lead to an explosion of solutions, to be developed, maintained, and kept consistent; meanwhile, Web hosting companies will be subject to growing service demands and will be lacking the technical man-power required to master them. With these premises, the strength of the W3I3[1] tool suite is to propose a model-driven approach to Web site design. Such an approach is based on WebML, a high-level language for specifying the structure of the content of a Web application and the organization and presentation of such a content in a Web site. In this paper, after a brief presentation of WebML, we concentrate on the W3I3 tool architecture, shown at work on case-study based on the popular site http://www.softseek.com.

1 Introduction

Designing data-intensive Web sites, i.e. sites whose primary purpose is the publishing of large volumes of data, is a primary concern for many companies. This challenge is going to become more demanding in the close future, because the activity of designing, deploying and evolving sites will face the need of serving content simultaneously to a variety of individuals or user groups, possibly equipped with different devices, each one characterized by specific rendition capabilities. In particular, WAP-compliant cellular phones, featuring WML-enabled micro-browsers [14], are already spreading in the market.

The W3I3 tool suite addresses personalized and multi-device content deployment by leveraging three different aspects of its architecture:

1. The possibility of organizing content at a high-level, using the WebML conceptual site modeling language ([5], http://webml.org). Alternative forms of content composition can be expressed as **site views**, and each site view

---

[1] W3I3 (Web-Based Intelligent Information Infrastructures) is a project funded by the EC, involving four companies and one Academic Institution (Politecnico di Milano) from four European countries

S.W. Liddle, H.C. Mayr, B. Thalheim (Eds.): ER 2000 Workshop, LNCS 1921, pp. 64–75, 2000.

may cluster information and services at the granularity most suitable to a particular class of users and devices.

2. The availability of an abstract presentation language, by which it is possible to construct reusable page descriptions (called **style sheets**) independent of the specific markup language required by the user's device. Style sheets specify pages in terms of content elements arranged in a nested grid model. They are written in XML [15].

3. The XSL-enabled translation technology [16,17], which maps abstract XML page specifications into concrete code in the languages of choice. The choice of language regards both the presentation, in which a specific markup language is selected (e.g., WML), and the binding of data to pages, where alternative server-side scripting languages can be used (e.g, Microsoft's Active Server Pages).

In this paper, after a brief presentation of WebML, we focus on the W3I3 tool suite architecture and on its individual components; we next show the tools at work in the modeling of an existing Web site (http://www.softseek.com).

2   The WebML Site Specification Language

WebML [5] is a high-level specification language allowing designers to express the core features of a site and abstracting them from architectural details. WebML concepts are represented in an intuitive graphic fashion, which can be easily supported by CASE tools and is conceived for non-technical members of the site development team (e.g., graphic designers and content producers). WebML internally relies on an XML syntax, which can be fed into software generators for automatically producing the implementation of a Web site. The specification of a site in WebML addresses four orthogonal perspectives: the structural model, the hypertext model, the presentation model, and the personalization model.

2.1   Structural Model

WebML does not propose yet another language for data modelling, but is compatible with classical notations like the E/R model [6], the ODMG object-oriented model [4], and UML class diagrams [3]. The fundamental elements of the WebML structural model are *entities* - acting as containers of data elements - and *relationships* - enabling the semantic association between entities. Entities have named properties, called *attributes*, with an associated type; properties with multiple occurrences can be represented by means of *multi-valued components*, which express a part-of relationship. Additional classical ingredients of conceptual models are present in WebML: *generalization hierarchies* for entities and *cardinality constraints* for relationships. An example of structural model for the SoftSeek case study is described in Section 4 and shown in Figure 2.

2.2   Hypertext Model

The hypertext model includes suitable constructs for representing one or more hypertexts, which can be published on top of the information described by the structure model. Each different hypertext defines a so-called site view; **site view** descriptions in turn consist of two sub-models, which are respectively called **composition** and **navigation** models.The composition model specifies which pages form the hypertext, and which content units (the atomic information elements that may appear in the Web site) make up a page. WebML content units are: data, multi-data, index, filter, scroller and direct units. Data units are used to publish the information of a single object (e.g., a software item), whereas the remaining types of units represent alternative ways to browse a set of objects (e.g., by presenting a subset of them in the same page, or by presenting an index, a search filter, first/last/previous/next scrolling commands, or finally by giving a direct access to a specific single element). Composition units are mapped to entities or relationships of the structural schema, from which they draw their content. The navigation model expresses how pages and content units are linked to form a hypertext. Links are either non-contextual, when they connect semantically independent pages (e.g., the page of an article to the home page of the site), or contextual, when the content of the destination unit of the link depends on the content of the source unit (e.g., the list of download sites associated to a given software item). Contextual links conform to the structure schema, because they connect content units whose underlying entities are associated by relationships in the structure schema. Examples of hypertext model for the SoftSeek case study are shown in Figure 3 and 4.

2.3   WebML Presentation Languages

Presentation is the modeling perspective concerned with the appearance of pages on the screen. WebML specifies presentation at the conceptual level, i.e., independently of the particular instance to be presented and on the specific rendition language. The basic unit of presentation is the **page**, as defined in composition modeling. Each page is associated to one or more **style sheets**, each specifying a different way of presenting its instances on the screen. Style sheets are XML documents obeying the WebML presentation DTD, which can be defined visually by means of a tool called Presentation Designer. The WebML presentation DTD includes tags for layout and content modeling.

The layout of each style sheet is a bi-dimensional rectangular space (represented by element space2d shown in Figure 1), which may include a set of possibly overlapping regions. Each region can be organized into a grid having an arbitrary number of rows and columns; each cell of the grid can recursively contain other regions, which can in turn be organized as nested grids. Cells of a grid are defined as the intersection of row and column ranges; therefore they may correspond to macro-cells made of several elementary cells forming a rectangular area. After layout definition, the next step in style sheet definition is to specify which piece

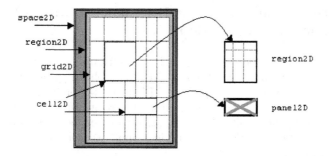

Fig. 1. Main XML elements pf WebML presentation DTD

of content goes into the various regions placed at the bottom of the hierarchical structure of the page (typically those contained in the cells of some grid). Content elements are specified using panels, i.e. XML fragments specifying an atomic or composite content element (e.g., a piece of text), which can be inserted into a region.

An example of page presentation applied to the running case of the SoftSeek site is detailed in Section 4 and shown in Figure 5.

Various rendition languages (HTML 3.2, HTML 4, WML, etc.) can be used to concretely implement the abstract presentation of pages. WebML presentation model can be extended to represent the peculiar aspects of a rendition language by including additional language-specific properties or constraints, which affect layout elements and panel templates. These extensions are again expressed in XML and collected in a document called *language profile*. For instance, a "background image" property for such layout elements as grids and cells is part of the language profile for HTML, but is not available in the profile for WML.

2.4   Personalization Model

Users and user groups are explicitly modelled in the structure schema in the form of predefined entities called User and Group (automatically provided by the site design tools, once a new project is created). The features of these entities can be used for storing group-specific or individual content, like shopping suggestions, list of favourites, and resources for graphic customisation, which can be published in a site view as normal content. In addition, OQL-like declarative expressions can be added to the structure schema, to define derived content based on the profile data stored in the User and Group entities, e.g., a discounted price based on the user's shopping history. Moreover, business rules can be generated in order to change the site's content according to user-specific information or to compute user profile data.

3   A Tool Architecture for Personalised, Multi-device Application Generation

WebML conceptual modelling is backed by a software architecture, which supports all the steps necessary to transform the WebML specification of a site into a running application in spectrum of industrial-strength rendition languages and server-side application platforms. The architecture of the WebML tool suite consists of three software layers: (a) the design tools, used for collecting design specifications (b) the device-specific template generators (c) the platform-specific adapters.

- The **design tools** support the modeling of Web applications at the conceptual level; the output of the design layer is a **conceptual schema** of the application, coded in WebML.
- The **code generation** layer transforms the conceptual schema of the application into an intermediate representation suitable for processing on top of commercial Web-database systems. This intermediate representation consists of a collection of **page templates**, which embody the structure, navigation, composition, and presentation of the application, but do not include to the actual data. Page templates are bound to a specific delivery language (e.g., HTM 3.2 or WML), and to a specific scripting language to be interpreted at server side (presently, MicroSoft's Active Server Pages, and JavaSoft's Java Server Pages).
- The **run-time adapters** consists of a set of lightweight Java components installed at the server-side, which give access to the actual data structures, which host the content of the entities and relationships defined at design time. This software layer shields the generated templates from the query language needed to bind the actual data to page when serving user requests. Presently W3I3 includes runtime adapters for wrapping JDBC compliant relational DBMSs and LDAP repositories.

3.1   Design Environment

The Design Environment includes three tools:

- **Site Designer**: permits the designer to define the structural model and the hypertext model of the application. Complex functions like the specification of derived data or the creation of an initial default site view are simplified by wizards, which, for instance, allow users to write OQL-like queries for expressing derived data in a visual way.the tool implements an advanced functions for user support, which perform the syntactic and semantic check of the project graph; if mistakes are detected, warning messages and tips are automatically presented to the user, which explain what is wrong and how to fix the problems
- **Presentation Designer**: deals with the specific aspect of presentation specification.The designer may define both generic and page-specific layouts. A

generic style sheet (also called presentation model, or untyped style sheet) is a specification of a page in terms of layout and fixed content elements (e.g., logos, fixed texts or images), which are independent of the specific objects used to fill the page. A **page-specific style sheet** (also called typed style sheet), instead, describes a page layout at a more detailed level, mentioning the actual elements (data fields, outgoing links, indexes, search forms, and so on) included in a certain page. Presentation Designer includes the support of multiple languages also.

– **Site Manager**: supplies all the required functions for publishing a W3I3 site on top of the runtime layer and data sources, and for maintaining it. These features are **site creation**,which invokes the Template Generator (see below) which builds the page templates necessary to run the application. the **site publishing** function is used to move all the application resources to the deployment server. The **user management** function addresses the specification of the access rights. Finally, Site Manager includes a **mapping function** for declaring the association between the structural model concepts and the repository structures chosen for the storage of data

The above tools are integrated by means of a further component, called **Repository Manager**, which manages the communications with all clients co-ordinate their access to the Central Design Repository, which hosts the WebML specifications in the form of XML documents and the graphic resources used in style sheets.

3.2   Template Generation

The Template Generator transforms a style sheet into a give rendition language (presently, HTML3.2 and WML are supported). The first use of the Template Generator is at design time to obtain a preview of the style sheet under construction. A **preview function** (launched from Presentation Designer) processes the XML specification of the style sheets, fetches the page characteristics from the design repository, and outputs a static file in the markup language of choice, in which the data content of the page is mocked-up (e.g., the value of attributes of type image are replaced by a reference to a constant image file). The second use of the Template Generator is at publication time, when the pages and style sheets of the site are transformed into ASP or JSP templates including instructions in a server-side scripting language for accessing the real data from the runtime data sources.

The Template Generator implements a multi-step process for transforming a WebML style sheet into a page template in a specific mark-up and server-side scripting language pair. The translation proceeds according to the following steps:

1. **Unfolding**: the original style sheet may contain composite panels and references to sub-pages, whose layout is described in a separate style sheet. The unfolding phase fetches all the necessary panel template and style sheets

definitions, and recursively replaces composite panels and sub-pages with their layout specification. The result is an unfolded style sheet equivalent to the original one but consisting only of atomic panels (images, texts, and anchors).

2. **Layout annotation** and optimisation: the unfolded style sheet is traversed to compute auxiliary information (e.g., the coordinates of the starting and ending point of all cells) and to apply optimisation operators to the unfolded layout (e.g., the removal of unnecessary nesting levels introduced by the procedure for recursive unfolding). At this stage, the process is still independent of the mark-up and server side scripting language.

3. **Data reference** translation: abstract data references contained in panels are converted from the WebML syntax to the syntax of the chosen server-side scripting language (e.g., Visual Basic Script). At this stage, the partially translated style sheet is bound to a specific server side platform.

4. **Mark-up translation**: the partially translated style sheet is fed to an XSL processor, which applies to it a set of rules, contained in an XSL file designed for the specific output mark-up language. The XSL file includes templates for mapping the WebML abstract layout tags into the most suitable constructs of the chosen rendition language. The output of mark-up translation is the final template, ready to be installed in the deployment server.

The core benefit of the described architecture is flexibility: a new rendition language can be easily added without changing the implementation. All the language dependent features are expressed in an XML-based syntax and the only tasks to be performed for integrating new languages are the creation of a new language profile (which requires the editing on an XML file), the creation or extension of panel templates to introduce language-specific properties (XML-based too) and the addition of an XSL file giving the rules for the mark-up conversion.

4   The SoftSeek Case-Study

As an example of WebML-driven site design, we now show how the popular Soft-Seek Web Site (http://www.softseek.com) can be modelled using the WebML tools, and re-engineered to obtain a version of the same content accessible via a WAP-enabled mobile phone. The SoftSeek Web Site allows searching, downloading and accessing documentation about software products. The software items are classified into several groups (editor's picks, top downloads, spotlight products, new releases, and so on) and are clustered in categories; each category has a name and a brief textual description, and can contain further sub- and sub-sub-categories. Categories and subcategories relate to spotlight and top products, whereas the sub-sub-categories include the complete listing of all products featured in that sub-sub-category. Each product is characterized by a set of technical data (e.g., version, size, release date, sample screenshot, descriptive text etc..), and is connected to the product's supplier, to related products from the same author, and to a set of download sites. Figure 2 shows the information published in the SoftSeek site represented as entities (categories, software items,

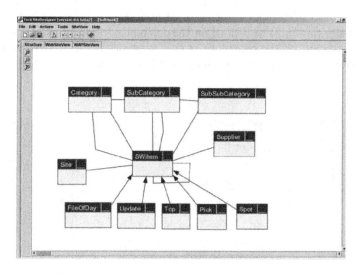

Fig. 2. Structural model of the case study edited with Site Designer

suppliers, download sites) and relationships (categories to subcategories, software items to download sites, suppliers and software items of the same supplier). An inheritance hierarchy represents classification of special software items.

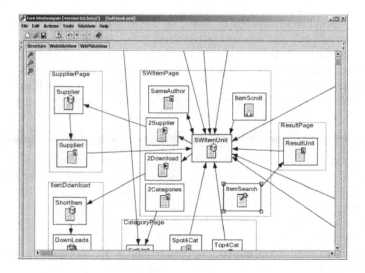

Fig. 3. An excerpt from the SoftSeek Web site view modeled in Site Designer

After consolidating the structural model, the hypertext model is designed: a different site view must be defined for the different devices,to cluster information differently based on the capability of each medium. Figure 3 shows a portion of the site view for the web version of the SoftSeek application. Due to space limitation,

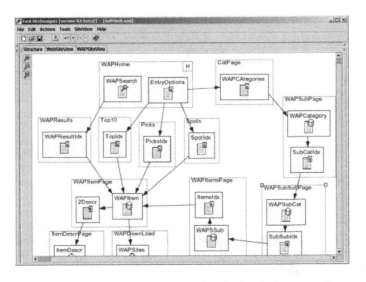

Fig. 4. Example of WAP site view for the SoftSeek case study

it is impossible to describe the complete schema and thus we concentrate on the design of the most representative pages only. At the top of the site view diagram, page *SWItemPage* describes the core information of a product.

Its center is the *SWItemUnit* data unit, which holds the product data (name, version, size, description text, image, and so on). *SWItemUnit* is linked via two direct units to the single supplier of the product (unit *Supplier* in *SupplierPage*), and to an alternative product page (*ItemDownload*), which includes a short description of the product (data unit *ShortItem*) and the set of sites wherefrom the product's file can be downloaded (multidata unit *Downloads*). *SWItemUnit* is also linked to the index of the other products by the same author (unit *SameAuthor*), from which it is possible to move to the page of another item, and to a scroller unit (*ItemScroll*), which permits the reader to move to the previous and next item in the same sub-sub-category. *SWItemPage* also contains an index of all the top-level categories (the index unit *2Categories*) and a search form (unit *ItemSearch*) to locate a product by keyword.

As an example of the support offered by the W3I3 tool suite to the specification and management of multi-device applications, figure 4 shows a second site view constructed on the same structure model of Figure 2, but aimed at

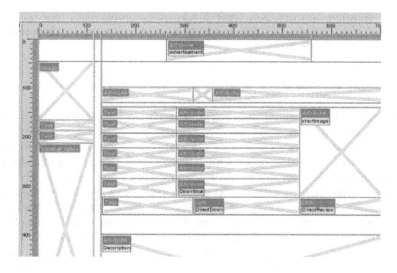

Fig. 5. Style sheet for the SWItemPage constructed in Presentation Designer

WAP devices. The WAP view is characterized by a finer granularity and less navigation options than the Web view, because wireless devices have a smaller display and thus entities must be split on different pages, and only the essential navigation facilities can be effectively used. Finally, we show the last step of application design in W3I3, which requires the use of Presentation Designer to define style sheets for the site view pages. Two sets of style sheets are required, one for PC browsers and one for WAP devices. Figure 5 shows a screenshot from the construction of the layout of the Web *SWItemPage*: different atomic and composite panels embedding the content elements of the page are arranged in a grid-based layout. The page was generated as so to be very similar to

Fig. 6. Output of the WML Template Generator on WAPItemPage

the actual page for showing software items in the softseek application(e.g. see www.softseek.com/Programming/C/Review 32546 index.html). Figure 6 show a wml page, obtained by running the Template Generator on a styel sheet edited in Presentation Designer using the WML language profile. Both page templates contain mark-up and sever side scripting code generated in a totally automatic way by the W3I3 tools.

5   Related Work

An array of tools is developed by the industry and academia to create and manage data-intensive Web sites. For brevity, we only present a selection of products that leverage some form of model-driven design, while we refer to [9] for a complete survey. **Designer 2000** [10] is a CASE environment included into the Oracle platform for deploying Web application based on a Web-enhanced entity-relationship design. Its Web Generator delivers PL/SQL code, which runs within the Oracle Web Server to produce HTML pages. Designer 2000 adopts a very database-centric approach while the W3I3 architecture adopts a mix of data, hypertext, and presentation modelling. In **Strudel**, a research project developed at AT&T Labs [7], Web sites are created from the declarative specification of the site's structure and content, in the form of queries over a data model for semi-structural information.**Homer** [2] is a CASE tool for building and maintaining data intensive Web site developed by researches of Università di Roma tre. The site schema is described by a formal model called AMD [1], which mixes database and hypertext concepts. Like the W3I3 tool suite, Homer can generate output for HTML and XML/XSL format, but Homer does not address presentation design. In the field of multi-device site development, Oracle's **Portal-To-Go**, [12] is a new server product that enables any existing database and Internet application to be made accessible from WAP phones, PDAs and other mobile devices. Portal-to-Go makes existing Internet or database applications device-independent, by extracting their content, dynamically converting it to XML, and then to the mark-up language supported by the user's device, including WML, TinyHTML, and VoxML. Portal-to-go, unlike W3I3, does not rely on a model-driven approach, and could be used as a runtime layer from the W3I3 tool suite, by adding an XSL translator targeting the Portal-To-Go XML DTD.

6   Conclusions

W3I3 tools have been applied to the modeling of a large number of case studies and applications, both in the context of the user companies of the W3I3 project, and by graduate students of our department. The W3I3 approach guarantees the following advantages:

- Increased productivity of Web developers. The use of a high-level, model-driven approach coupled to fully automatic code generation facilitates the design and thus lowers the technical edge of Web developers, alleviating an essential problem of most Web hosting companies.
- Lower ownership costs. The use of a model-driven approach eases mainten-ance and evolution, because changes can be analyzed at a higher level and

propagated to the implementation. This feature is essential in view of the exponential growth of complexity of Web sites caused by the interplay of multi-device output and one-to-one delivery.

– Higher consistency across applications delivered to different output devices. We envision Web applications that offer a consistent and stateful interaction to users who, e.g., initially connect from home (e.g., via digital TV), then access the application while traveling (by means of a mobile device), then in the office (via personal computer). Interaction uniformity is guaranteed by W3I3's content-centric design, where information is modeled once and then adapted to different media and deployed automatically.

All the key features of the W3I3 tools are fully implemented, and the final version of the W3I3 tool suite will be available by the end of the W3I3 project (October 2000).

References

1. P.Atzeni, G. Mecca, P. Merialdo, To Weave the WEB, VLDB 1997
2. P. Merialdo, P. Atzeni, M. Magnante, G. Mecca, M. Pecorone: Homer: a Model-Based CASE Tool for Data-Intensive Web Sites. SIGMOD Conference 2000
3. G. Booch, I. Jacobson, and J. Rumbaugh, The Unified Modeling Language User Guide, The Addison-Wesley Object Technology Series, 1998.
4. R. G. G. Cattell, Douglas K. Barry, and Dirk Bartels (Eds.), The Object Database Standard: ODMG 2.0, Morgan-Kaufmann Series in Data Management Systems, 1997
5. Ceri, Fraternali, Bongio, Web Modeling Language (WebML): a modeling language for designing Web sites, WWW9 conference, Amsterdam, 15-19 May 2000
6. P. P. Chen, The Entity-Relationship Model, Towards a Unified View of Data, ACM-Transactions on Database Systems, 1:1, 1976, pp. 9-36.
7. Fernandez, Florescu, Levy, Suciu, Catching the boat with Strudel: Experiences with a Web-Site Management System. In Haas and Tiwary eds, Proc. Int. Conf, Sigmod 1998
8. P, Fraternali, P. Paolini, Model-Driven Development of Web Applications: the Autoweb System, to appear into Transaction on Information System, 2000
9. P. Fraternali, Tools and Approaches for Developing Data Intensive Web Application: a Survey. ACM Comp. Surveys, Volume 31, No. 3 (Sep. 1999),
10. Oracle Designer/2000 WebServer Generator Technical Overview, Oracle Cor.
11. Lightweight Directory Access Protocol (v3), http://www.cis.ohio-state.edu/htbin/rfc/rfc2251.html , 1997
12. Oracle portal-to-go http://www.oracle.com/mobile/portaltogo/index.html
13. M. Abrams, C. Phanoriou et. al.: UIML: an Appliance-independent XML User Interface Language, Proc. WWW8, Elsevier, pp. 617-630.
14. Wireless markup language http://www.wapforum.org, Wap Forum Ltd.
15. Extensible Markup Language (XML) 1.0, http://www.w3.org/TR/1998/REC-xml-19980210, 1998,
16. Extensible Stylesheet Language (XSL) 1.0, http://www.w3.org/TR/xsl/, 2000
17. XSL Transformations (XSLT) Version 1.0, http://www.w3.org/TR/xslt, 1999

Abstraction and Reuse Mechanisms in Web Application Models

Gustavo Rossi[1], Daniel Schwabe[2], and Fernando Lyardet[1]

[1]LIFIA Facultad de Informática. UNLP.
La Plata, Argentina
{gustavo,fer}@sol.info.unlp.edu.ar

[2]Departamento de Informática, PUC-Rio, Brazil
schwabe@inf.puc-rio.br

Abstract. In this paper we analyze different abstraction and reuse mechanisms that should be used in Web applications to improve their evolution and maintenance. We first review the OOHDM approach for defining a Web application model, in particular the separation of the navigational model from the conceptual model. We next focus on abstraction and composition mechanisms in both models showing how to combine OOHDM's views with the concept of node aggregation. We introduce navigation and interface patterns and show the way in which patterns generate the architecture of Web design frameworks. We strongly argue that in the currently state of the art of Web applications we can build models of families of similar applications to improve design reuse. Next, we present our notation for specifying Web frameworks, giving some examples in the field of E-commerce. Some further work is finally discussed.

Introduction

Building complex Web applications is a time consuming task as they must provide navigational access to critical information resources, not only allowing the user to browse through the potentially large universe of information but also to operate on it. In some areas such as electronic commerce, customers' actions trigger a sophisticated workflow that must be integrated with the core business software. The first obvious consequence is that we must not only design the navigational architecture carefully but also integrate it effectively with the underlying business model.

To complicate matters, Web applications should be developed with zero defects, with short deployment and maintenance times. In this context, we should use not only systematic engineering techniques but also be able to improve reuse during the whole development cycle. The key for obtaining reusable designs or components is to be able to build extensible and reusable conceptual models. However, while reuse techniques have been widely explored for conventional applications [6], the very nature of Web applications seems to prevent designers from being able to cope with design and implementation reuse.

S.W. Liddle, H.C. Mayr, B. Thalheim (Eds.): ER 2000 Workshop, LNCS 1921, pp. 76-88, 2000.

The purpose of this paper is to present different design reuse mechanisms that should be used while building Web application models. We stress mechanisms such as navigation patterns and Web frameworks, particularly those that apply to Web applications, such as contexts. In this sense, our goal is not to present novel design primitives, though we introduce some, like aggregates and generic contexts; we rather seek to motivate discussion on the problem of reuse in Web application models.

Though we use OOHDM [7, 10, 11] as the base design method, the ideas in this paper can be easily applied to other modeling approaches. In section 2 we characterize Web application models as the combination of conceptual and navigational models. In section 3 we show how different abstraction and composition mechanisms in OOHDM work together to achieve elegant and reusable design models. In section 4, we briefly address abstract design reuse by reviewing navigation patterns. Since patterns generate architectures, we go further in section 5 and present Web design frameworks as a way to achieve reuse of entire domain models. In section 6 we present OOHDM-Frame, a notation for specifying Web design frameworks. Some further work is finally discussed.

Web Application Models: Conceptual + Navigation Models

The key concept in OOHDM is that Web application models involve a Conceptual and a Navigational Model [10]. The conceptual model aims at capturing the domain semantics using well-known object-oriented primitives and abstraction mechanisms. In an electronic store for example, the conceptual model will contain core classes such as Product, Order, Customer, etc. with their corresponding behaviors. We use UML as the notation to specify the conceptual model. Since the conceptual model is an object-oriented model, we can use existing reuse approaches in object-orientation [1, 6].

In the OOHDM approach the user does not navigate conceptual objects but navigation objects (nodes). Nodes are defined as views on conceptual objects, using a language that is similar to OODB view-definition approaches [5]. Nodes are complemented with links that are themselves specified as views on conceptual relationships. The navigational schema shows the node and link classes that comprise the navigational structure of the application. For each particular user profile we build a navigational model as a view of the shared conceptual model. In this way, we can reuse the conceptual model in a family of similar applications. Moreover, as shown in section 3, we can define different views in the context of a single application.

In Fig. 1 we show part of the conceptual model of an electronic store. Notice that some classes in the model will be mapped onto the navigational model (i.e. they will be explored as nodes) while others, such as PaymentMethod, will not.

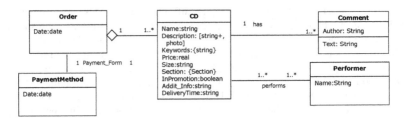

Fig. 1. Conceptual Model of CD store.

If we are designing the customer view of the electronic store, we will specify node classes for products. As shown in Fig. 2, these nodes may combine some attributes of conceptual class CD with attributes from conceptual class Comments and Performer. Notice that in good object-oriented software specifications (such as the one in Figure 1), products, comments and performers belong to different classes -. Nodes meanwhile implement opportunistic views of conceptual classes (following the Observer design pattern [2]). The precise syntax for defining views can be found in [10].

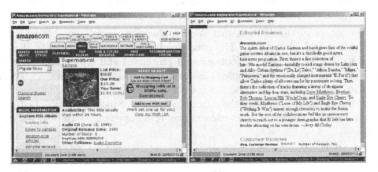

Node CD FROM CD:C
    name: String, price: Number
    performer: String SELECT name FROM Performer: P WHERE C *isPerformed by* P
    comments:Array[Text]   SELECT   text   FROM   Comment: R   WHERE
        C *hasComment* R
    other attributes and anchors

Fig. 2. CDs including comments in Amazon.com and the OOHDM definition.

The Navigational Schema is complemented in OOHDM with a Context Schema that shows the navigational contexts and access structures (indexes) in the application. A navigational context is a set of objects that are usually explored sequentially; for example: Books of an author, CDs by a rock band, etc. There are different kinds of navigational contexts: class derived, link derived, arbitrary, etc [11]. Access Structures act as indexes to group of related objects; they are specified by indicating the target objects and the selector to be used in the index.

In Figure 3 we show part of the context schema for the electronic store. The notation in Figure 3 shows in a compact way, which sets the user will explore, and how they are related with each other. Navigational contexts are a novel design primitive for specifying sets in a concise way, specifically developed for exploring hyperspaces.

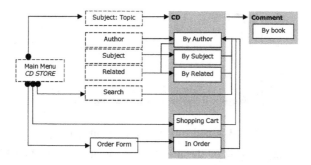

Fig. 3. Context Schema for the CD store.

Dashed boxes in Figure 3 show access structures (indexes) while boxes inside Class CD (and comment) indicate possible contexts in which a CD (respectively a comment) can be accessed. A node may appear in different contexts, showing different information according to the context within which it is reached. In this situation, we use Decorators [2] to decouple the base information in the node from the different "faces" this node exhibits. Consequently, navigational contexts combine two navigational patterns, Set-based navigation and Nodes in Context [9]. The navigational and the context schemas play an important role when reusing application models in a family of applications in the same domain. We will discuss this kind of reuse in section 6.

Combining Views with Aggregate Nodes

Complex Web applications provide multiple ways of reaching the information they contain. In e-commerce applications for example customers receive different kinds of advising such as hot-lists, recommendations, new releases, etc. In Figure 4 we show an example of a home page that contains different kind of links to products in an electronic store. In OOHDM we can aggregate nodes to specify this home page. An aggregate allows gluing different information items (other nodes) and access structures (like indexes) in the same node.

Fig. 4. A node representing a home page.

The specification of part of the node for the home page in Figure 4 reads as follows:

```
Node MusicHome
news: Array [CDView]
search: SearchTool
categories: IndexOfCategories
topSellers: IndexOfTop
landmarks: IndexOfStores
...
other attributes
```

```
Node CDView FROM CD: C
name: String
performer: String SELECT name FROM Performer: P WHERE C isPerformed by P
description: Photo
shortComment: Text
```

Notice that the specification of type CDView above takes profit of the viewing mechanism and it can be reused in other parts of the site (for example the Artists Essentials section uses a similar summary for each CD). Aggregates allow specifying composite nodes in an opportunistic way (as it is usual in Home pages). However, aggregate nodes combine with the viewing mechanism in a way that goes beyond simple composition mechanisms in object-orientation. This synergy is complemented with the linking mechanism that allows different views of the same object to be connected with each other. For example you can easily navigate from the summaries of CDs in Figure 4 to the corresponding CD. In Figure 5 we show in a diagram how to reuse one object's view and how this view is linked to another one of the same object.

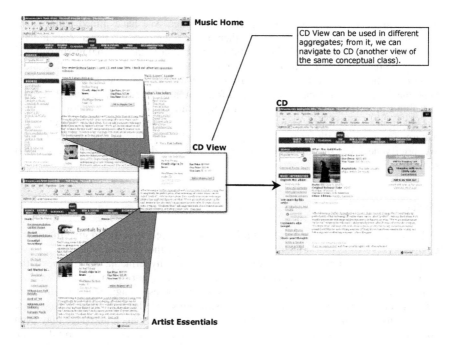

Fig. 5. Aggregates and view reuse in a navigational schema.

This simple example raises some interesting issues and questions related with design reuse:

1. Can we generalize the basic idea behind the previously shown home page? What design problem are we solving when building this kind of aggregate node? Can we apply this same solution in other Web applications?

2. Is the structure of this application similar to others in the same domain? In other words: how can we profit from our intellectual investment while designing the conceptual and navigational models in similar applications?

These questions show some non-trivial design reuse problems. While composition, viewing and inheritance allow improving reuse and maintenance in a single application, they are not enough for expressing reusable aspects in a family. We next introduce two novel approaches for design reuse in Web applications: navigation patterns and Web design frameworks.

Design Reuse Using Navigation Patterns

Patterns record design experience by expressing in an abstract way recurrent problems and proven solutions. They are a wonderful tool for capturing, conveying and reusing design experience.Patterns complement design methods by showing solutions that go beyond naive uses of the methods' primitives. Patterns improve communication among designers by enriching the design vocabulary with terms that express non-trivial design structures. They formalize well-known solutions in such a way that novice designers can profit form experts' knowledge. We have mined patterns for Web applications and have documented them using a template similar to Alexander's one [9]. In fact hypermedia and Web patterns are similar to the original urban patterns as they express recurrent structures for building usable navigable spaces; they show design solutions that help the user find his way through the hyperspace. The hypermedia community have proposed dozens of new patterns [3], and it is now pursuing a project for expressing these reusable solutions in a shared catalogue [4].

Continuing with the previous example we may define two simple but effective patterns for dealing with (part of) the application's complexity: Portal and Landmark. We briefly describe them, stating the problem they address and the (widely used) solution.

Portal

In many Web applications, particularly in E-commerce we want to give the user a comprehensive description about what he will find in the site including daily news, suggestions, opportunities, etc. If we follow a naive hypermedia design view, the "home" page should map some conceptual object, or may just be an index to services or products. The solution is to design the home (or homes) as aggregates of different information items, anchors and access structures, Dedicating space to news, suggestions to the user, general indexes, special offers, etc. This home page may even contain information that may not be "semantically" connected. A portal is an opportunistic design solution that allows increasing the site's number of visitors as it is easier and quicker for them to find what they want. Portals are widely used in all e-commerce sites such as amazon.com, netgrocer.com and more general sites such as netscape.com. Portals generalize the design solution in Figure 4.

Landmark

Many Web applications contain sub-sites that provide specific functionality (different shops, search facilities, etc). When we describe the navigational schema (i.e. the network of nodes and links types), we try to follow closely those relationships existing in the underlying object model; for example we can navigate from an author to his books, from a CD to the list of songs it includes. However, we may want that at any moment the reader can jump to the music or book (sub) stores or to his shopping basket. The solution is to define a set of landmarks and make them accessible from every node in the network. making the interface of links to a landmark look uniform. In this way users will have a consistent visual cue about the landmark. We may have different levels of landmarks according to the site area we are visiting. Landmarks are different from indexes as they appear in every node in the application. This pattern is widely used in Web applications for indicating relevant sub-sites and functionality.

Patterns do not stand by themselves. They must be integrated into the development method in order to be effective. They must be combined to create higher level abstractions. In the context of OOHDM we have defined notations for some navigation patterns such as Set-Based Navigation and Nodes in Context [10] and Landmarks [9]. In Figure 6 we generalize the preceding example by showing a navigation model incorporating the idea of Landmarks. Notice that instead of a tangled diagram we get a simplified one in which links to landmarks are omitted. CD Store, BookStore and Toy Store in Figure 6 are Landmarks (indicated with an arrow with a bullet as source). Notice that, within CD Store, "Subjects", "Search", "Shopping Cart" and "Order" are second level Landmarks.

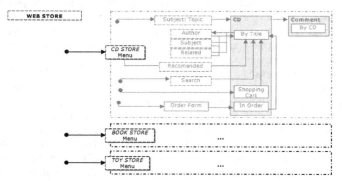

Fig. 6. Using Landmarks in the Navigational Schema.

Incorporating patterns into the design armory helps to reduce the complexity of diagrams thus making reuse more feasible. However, when we design complex applications we need more powerful reuse approaches.

In the e-commerce domain for example we can easily find that most virtual stores offer similar services to the customer: most of them allow finding products by searching or hierarchical navigation, all of them provide a shopping basket for making selections persistent, etc. Moreover we can find commonalties even in the core application behavior: for example, the set of actions triggered when a customer makes a check-out operation are basically identical: verifying user data, creating an order, sending a confirmation mail, sending another mail when products are shipped, etc. We should be able to define architectures that abstract these commonalties and that can be extended smoothly to cope with variations in each particular application. We next introduce Web design frameworks and show how they relate with navigation patterns.

From Web Patterns to Web Frameworks

Frameworks are reusable designs for a family of applications in a particular domain. They act as skeletons of a set of applications that can be customized by an application developer. When many different applications must be constructed in the same domain, application frameworks provide "templates" for supporting their commonalties, and accommodating individual variations (differences). While patterns provide abstract reuse of design experience, frameworks allow reusing concrete designs in a domain [1]. Frameworks are composed of a set of abstract and concrete classes, which contain the specification of generic behaviors (usually specified using a particular programming language) in the intended domain. A key aspect for designing frameworks is identifying its hot spots (i.e.: the points in the framework where variations will appear). Following with the preceding example, we can generalize the conceptual model (in Figure 4) to reflect abstract classes and collaborations in virtual stores. The model should include an abstract class Product, different kinds of Orders and Payment Methods, Comments, etc. A designer developing a particular store will need to define new concrete classes (for example sub-classes of Product) and specialize some behavior such as order processing, to accommodate it to the particular application (for example, selling other products using different business rules). In virtual stores (such as Amazon.com) the approach will work for defining new sub-stores in the company that may have, for example, different shipping or payment policies.

Designing frameworks is a difficult but rewarding task. We need to understand the domain and produce a generic design that can be instantiated into different applications. To apply this approach to Web application models, we need to take into account different kinds of variability: those related with the domain model (e.g. different payment policies) and those related with navigation architectures (e.g. different indexes, contexts, etc). Besides, programming-language-centric approaches (common in application frameworks) are difficult to apply in the Web, given the large number of combinations of languages and tools that are often used in Web application development and implementation.

We define a Web *design* framework as a generic design of possible Web application architectures, including conceptual, navigational and interface aspects, in a given domain. We have used the OOHDM model as the basis architecture for specifying Web design frameworks. Web design frameworks comprise a generic conceptual model (that may be itself an object-oriented framework), a generic navigation schema and a generic context schema.

Web *design* frameworks are different from *application* frameworks because while the latter are programmed in a specific language, Web *design* frameworks are environment and language-independent. Web design frameworks include an additional perspective with respect to conventional application frameworks: the generic navigation architecture.

Web design frameworks can be mapped either to an application framework to be later instantiated into a running application or can be instantiated into "pure" OOHDM models and then implemented as a single Web application [12]. We next present a notation for improving Web application models with the kind of abstractions needed in Web design frameworks.

OOHDM-Frame: A Notation for Web Frameworks

In order to specify Web design frameworks, we have defined a new notation, called OOHDM-Frame that extends OOHDM smoothly. It is not our objective in this paper to give the detailed syntax of the notation but rather to analyze how to improve existing abstraction and composition mechanisms in conceptual modeling in order to express generic Web functionality. We will present the notation briefly to stress each particular modeling feature. As previously explained, the specification of a framework's model in OOHDM-Frame is comprised of generic Conceptual and Navigational Models specifications, together with instantiation rules. We next analyze each one pointing out novel abstraction mechanisms.

Abstraction and Genericity in the Conceptual Model

Variability in Web applications may appear in the conceptual model. In Figure 7 we show part of a generic model for electronic stores. Notice that we have included some abstract classes like Product and specialized Comment and Payment Method.

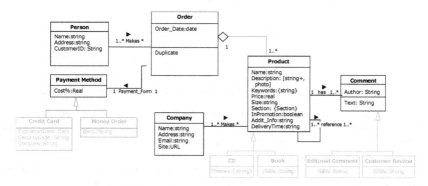

Fig. 7. A generic conceptual model for virtual stores.

Genericity in object-oriented models has been largely discussed in the object-oriented community and one can use existing notations to express generic classes and behaviors [8], so we don't discuss it further here.

Specifying Generic Navigational Models

A generic Navigation Model in OOHDM-Frame is made up of a Generic Navigation Schema, a Generic Context Diagram, and a set of mapping and instantiation rules. The Generic Navigation Schema generalizes the idea of views (or observations in the terminology of [2]). It is similar to the Navigation Schema, except for the fact that Node attributes may be optional (marked with an "*") and Relations (links) can be optional (drawn with a dashed line), as shown in Figure 8. An optional attribute (respectively Link) may or may not appear in an instantiated application. Notice that as the navigational model will be often mapped into a non object-oriented implementation, we are not constrained to "pure" notations, e.g. we can always specify optional features (attributes or links) by defining appropriate class hierarchies, though in a less concise way. For the sake of simplicity we have not included those sub-classes in Figure 8.

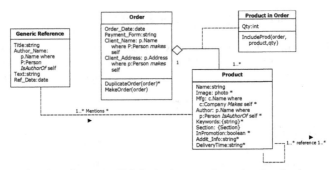

Fig. 8. Optional attributes and Links in the generic navigational schema

Sub classing in the Generic Navigational Schema allows a more subtle way of achieving genericity. In the example above, we may create a sub-class of Product and either add an attribute or anchor or we may even need to specialize the view specification for a particular attribute, as shown below.

Suppose for example that we have two sub-classes of Comment (as shown in the generic conceptual schema of Figure 7); if we want to generalize the store to a Books and CDs store (in the context of a framework for virtual stores), we may require that some of the navigational Product sub-classes show comments from only one (conceptual) sub-type. Accordingly, we show the specification of part of the abstract node class Product, and how we specialized the definition of the attribute *comments* for Books. The Refine operator takes the query in the corresponding super-class and replaces Comment with its sub-class EditorialComment. We are thus indicating that books only show Editorial Comments.

Node Product from Product: P
comments: Array[Text] SELECT text FROM Comment: R WHERE P *hasComment* R

Node Book from Book:B
REFINE comments WITH EditorialComments

Generic Context Diagrams meanwhile represent another kind of hot spot in Web applications, showing in an abstract way which contexts and access structures may appear in a particular domain. Notice that as navigational contexts are sets of nodes, defining generic contexts is equivalent to specifying generic sets. Thus, achieving generictiy in a context diagram is not straightforward with usual object-oriented abstraction mechanisms, i.e. though context and indexes may be finally mapped into classes, expressing their variability may require using complex diagrams. Instead, we preferred to generalize Context Diagrams and to complement them with a generic context specification card providing a guide for the implementers indicating possible restrictions. In Figure 9 we show a simplified generic Context Schema for our virtual store framework. Dashed boxes and rounded boxes indicate generic access structures and contexts. For example the generic context "Product by Property" will be typically instantiated into one or more contexts that allow navigation among products according to certain properties (e.g., "Product by price"; "Product by author"; etc…). Once within any of these, it is normally possible to navigate to other "Related Products" (e.g., accessories, matching products, etc…). There are several access structures that lead the reader into these contexts; typically, these are hierarchical access structures that reflect product sections (departments) in a real world store. Notice that we have also specified some Landmarks (like Shopping Card, Order Form and Search). A second way to look at products is within arbitrary groupings obtained opportunistically. Typically, these will correspond to some person's (or publication) recommendations, or some guide, such as "N.Y. Times Bestsellers List". This grouping is modeled through the generic context "Products by Reference".

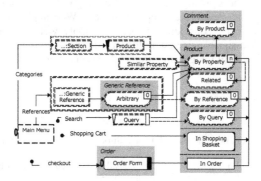

Fig. 9. Generic Context Schema for virtual stores.

The context diagram in Figure 3 is an instantiation of the generic diagram in Figure 9, where the generic context "Product by Property" has been concretized into "CD by Subject" and "CD by Author". Generic Context Schemas show concisely different ways of providing Set-based navigation in Web applications for a particular domain.

Concluding Remarks and Further Work

We have discussed in this paper different abstraction and reuse mechanisms in the context of Web applications. We have shown that even the simplest techniques like composition and inheritance offer subtle combinations to the designer when dealing with non-trivial navigation models. In particular, the OOHDM viewing language can be used synergistically with aggregation (and sub-classification) to produce compact and reusable navigation designs. We have discussed reuse of design experience by briefly analyzing navigation patterns. Although patterns provide design reuse at a fine granularity, we have shown how to combine them to obtain larger reusable models. We have introduced Web design frameworks, explaining how generic and reusable conceptual and navigational models can be described using the OOHDM-Frame notation. Web design frameworks show how the combination of patterns (like Set-Based Navigation, Landmark, and Observer) may yield a generic design for a family of applications in a particular domain.

Even though the focus of this paper has been put on design, it is important to stress that all primitives and mechanisms previously presented can be implemented using current Web technologies [12]; in addition, mapping design frameworks to "pure" object-oriented settings is straightforward. We are mining Web patterns in specific domains such as e-commerce, and studying ways to enrich the framework design notation with these new patterns. Several implementation aspects should also still be studied, such as efficient ways to implement views and contexts in Web applications. Another aspect that we did not address is the use of support tools both for drawing diagrams and generating code. While UML tools like Rational Rose can be used we still have to build similar editors that support the slight syntactic and semantic differences among UML and OOHDM.

We believe that the growing interest in Web applications requires ways to build easily extendable and reusable conceptual models. Web applications present novel features that need to be considered in order to apply well-known abstraction and composition mechanisms to this new field. The ideas underlying this paper may serve as the background for studying abstraction and reuse in Web models.

References

1.   M. Fayad, D. Schmidt and R. Johnson (editors): "Building Application Frameworks", Wiley 1999.
2.   E. Gamma, R. Helm, R. Johnson and J. Vlissides: "Design Patterns. Elements of reusable object-oriented software". Addison Wesley, 1995.
3.   F. Garzotto, P. Paolini, D. Bolchini and S. Valenti: "Modelling by patterns of Web applications". Proceedings of the First International Workshop on Conceptual Modeling and the WWW, Paris, France, November 1999,Lecture Notes in Computer Science, Vol. 1727, Springer, 1999, 293-306.
4.   Hypermedia Patterns repository: http://www.designpattern.lu.unisi.ch.
5.   W. Kim, "Advanced Database systems", ACM Press, 1994.
6.   Bertrand Meyer, "Reusable Software" - The base object-oriented component libraries. Prentice Hall 1994.
7.   Daniel Schwabe and Patricia Vilain: "The OOHDM notation", available at http://sol.info.unlp.edu.ar/notacaoOOHDM/
8.   W. Pree: "Design Patterns for object-oriented software", Addison Wesley, 1994.
9.   G. Rossi, F. Lyardet and D. Schwabe: "Patterns for designing navigable spaces" Pattern Languages of Programs 4, Addison Wesley, 1999.
10. G. Rossi, D. Schwabe, F. Lyardet: "Web application models are more than conceptual models". Proceedings of the First International Workshop on Conceptual Modeling and the WWW, Paris, France, November 1999, Lecture Notes in Computer Science, Vol. 1727, Springer, 1999, 239-253.
11. D. Schwabe, G. Rossi: "An object-oriented approach to web-based application design". Theory and Practice of Object Systems (TAPOS), Special Issue on the Internet, v. 4#4, pp.207-225, October, 1998.
12. D. Schwabe, G. Rossi, L. Emeraldo, F. Lyardet: "Web Design Frameworks: An approach to improve reuse in Web applications. Proceedings of the WWW9 Web Engineering Workshop, Springer Verlag LNCS, forthcoming.

From Web Sites to Web Applications:
New Issues for Conceptual Modeling

Luciano Baresi, Franca Garzotto, and Paolo Paolini

Dipartimento di Elettronica e Informazione - Politecnico di Milano
Piazza Leonardo da Vinci 32, 20133 Milano, Italy
{baresi, garzotto, paolini}@elet.polimi.it
Phone: +39-02-2399 3638; fax: +39-02-2399 3411

Abstract E-commerce, web-based booking systems, and on-line auction systems are only a few examples that demonstrate how web sites are evolving from
essentially read-only information repositories to distributed applications. These
new *web applications* blend navigation and browsing capabilities, common to
hypermedia, with "classical" operations (or transactions), common to traditional
information systems. The coupling between hypermedia and operational features raises a number of novel modeling issues. Conceptual modeling for web
applications is not just the union of two activities performed in isolation - designing the operations and designing the hypermedia aspects. Rather, modeling
the *integration* (and *interference) of the two facets of design* (hypermedia and
operations) is the issue. The co-existence of operational and navigational aspects
poses several new problems to designers: For example, how do information
structures and navigation support operations? How do operations affect information structures and navigation? How do operations and navigation interplay?
How are user tasks related to both navigation and operations?
The paper discusses these and other questions, and provides a contribution to-
ward possible solutions, based upon the W2000 design framework.

1. Introduction

The today prominent shift of web sites from essentially read-only information repositories to real *applications* running on the web introduces a number of challenging issues in web design ([7]).

In this paper, we mainly concentrate on the modeling problems raised by two
"new" features of web applications: the extended role of "operations", which comple-
ments navigation with typical information system functionality (e.g., business *transactions*), and the *tight integration* among these different kinds of "operations", that
mutually interfere and induce complex interaction mechanisms. For example, if we
considered a "typical" e-commerce web site for buying music, users, while navigating
within the catalog of records, can filter the products of interest by formulating a query,
send a message to the virtual shop to either transmit a specific request or send a com-
plaint. But, they can also bookmark some products and include them in their "shop

S.W. Liddle, H.C. Mayr, B. Thalheim (Eds.): ER 2000 Workshop, LNCS 1921, pp. 89-100, 2000.

ping bags", inspect the chosen products by navigating from the bag to selected records, exclude some of the previously chosen items, evaluate the total cost of the final list, and, after providing additional information, complete the buying transaction.

The problem of conceptual design of web applications can be approached from two different perspectives:
- Web applications can be regarded as an extension to the traditional information systems, complemented with navigation and complex information structures;
- Web applications can be regarded as an extension to traditional web sites complemented with various kinds of application operations and conventional functionality of information systems.

Whatever the chosen approach, we claim that conceptual modeling of a web application is not just the union of two tasks performed in isolation: modeling typical hypermedia aspects and modeling operations. It sums up the complexity of the two activities with the additional task of designing their coupling. As such, conceptual modeling of web applications cannot simply pile up existing techniques of hypermedia design (borrowed from the hypermedia/web communities) with methods and notations for "traditional" operation modeling (borrowed from the information systems or software engineering communities). The crucial point is to integrate and extend models, design methodologies, and techniques to meet the new design challenge.

This paper will discuss the above issues, will analyze some examples, and will sketch some methodological solutions adopted in the W2000 design framework. W2000 tackle the problem of designing web applications regarding them as extensions of traditional web sites. It borrows hypermedia design concepts from HDM2000 [2], the latest version of the Hypertext Design Model [13, 14], and extends them to model the peculiar dynamic features of web applications, i.e., operations and their intertwining with navigation and hypermedia aspects. W2000 primitives are specified using suitable extensions of the Unified Modeling Language UML [5, 8, 9]. The advantage of UML [5, 8, 9] is to provide a well known, standardized, and customizable graphical notation that is supported by a number of existing case tools.

The rest of this paper is organized as follows. Section 2 identifies critical modeling requirements for web applications. Section 3 discusses how to blend the design of hypermedia structures with user operations. Section 4 describes how to model evolution in web applications. Section 5 discusses how to couple operations' behavior with hypermedia. Section 6 concludes the paper. Lack of space prevents us from explaining all details of W2000.

2. Modeling Requirements for Web Applications

Traditional web sites can be regarded as (universally distributed) hypermedia applications and can largely be modeled using hypermedia modeling methods. Most approaches [1, 3, 6, 11, 12, 15, 17, 16] distinguish at least two main dimensions for hypermedia/web site conceptual modeling: information modeling and navigation modeling[1]. Information modeling describes the contents of the web site. Navigation modeling describes its navigation capabilities, i.e., the paths that users can traverse to explore the information universe. In this paper, the union of information modeling and navigation modeling will be globally referred to as *hypermedia modeling* or *web modeling*, indifferently.

Traditional hypermedia modeling focuses on organizing the information structures and navigation paths. Dynamic aspects are neglected, for two main reasons. Firstly, no *read* operations need to be modeled explicitly: the only user operation, *follow-a-link*, has a built-in behavior, which can be left implicit. Secondly, information and navigation structures are not expected to evolve at run time. Therefore operations that explicitly manipulate those structures are not necessary.

Web applications, instead, introduce a new dimension, which must be modeled explicitly: non-navigational *operations*. These operations are not "read-only": They may modify individual contents as well as entire information/navigation structures of the application. As such, they add *dynamic* (i.e., evolution) properties to the two original dimensions (information and navigation) of conventional web sites.

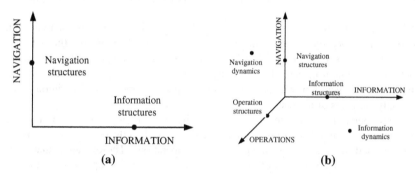

Figure 1: Design Spaces for Web Sites and Web Applications

Graphically, we can represent the design space for web sites as a 2D space, as shown in Figure 1(a): web site modeling includes only information and navigation *structures*. Figure 1(b) describes the 3D modeling space for web applications and points out what must be added to traditional hypermedia design: i) Operations must be modeled explicitly; ii) Operations may refer to both information and navigation structures; iii) Information and navigation are influenced by operations and thus their dynamics must be modeled explicitly.

3. Blending Hypermedia and Operations

Before discussing how to jointly specify hypermedia structures and operations, we need some basic notions and terminology borrowed from hypermedia modeling. Al-

---

[1] Various approaches explicitly include also presentation modeling, which we omit to discuss here since it is less affected by the introduction of operations.

though we refer to the web design model HDM2000 [2], most of the following concepts are absolutely general, that is, independent from the specific model being used.

The *hyperbase layer* contains the base information structures of the web application, together with the navigation paths to traverse them. It comprises:

- A set of *entities* that model all relevant domain objects. An entity can be structured in different parts, called *components*. Components are usually (but not necessarily) materialized as web pages interconnected by (structural) *links*. Entities can be either singleton objects or instances of *entity types*.
- A set of *semantic associations* that define the semantic *links* among entities. Semantic associations can be either singleton objects or instances of *semantic association types*.

The *access layer* is a means for facilitating the comprehension and accessibility of the hyperbase. It comprises *collections*, that is, containers (of entities, components, or other collections) that provide superimposed information ([10]) and alternative groupings for the hyperbase elements. If we considered a web-based conference manager, the hyperbase would contain several instances of the entity type *Paper* and *Accepted Papers* could be a smart example of collection, that is, a way for grouping related elements[2].

Moving to operations, we can organize web application operations in two main categories: *user operations* and *system operations*. User operations are the only operations directly visible to users. The *buy it* operation, for example, that allows users to put a selected item in their shopping bags, is a typical user operation. Another example is a filter operation, which allows users to specify some selection criteria to retrieve only elements of interest.

System operations are triggered by either navigation actions, or user operations, or maintenance (administrator) operations. They are not directly perceivable to users, but are needed to specify the semantics of their operations. The operation that, triggered by *buy it*, updates the list of items in the shopping bag, the operation that registers all links traversed by a given user for profile tracing, the query operation that applies user filtering operations, all are examples of system operations. System operations are at a lower abstraction level with respect to user operations. As such, we believe that they can be omitted during conceptual design, which should instead focus on user operations only. System operations can be deferred to a different phase, i.e., during detailed design, which is more implementation-oriented and requires that the actual behavior of operations be specified.

Once user operations have been identified, the designer must model the way they become available to users. This task is absolutely new with respect to traditional hypermedia design where operations are not first-class citizens, and raises some new modeling issues. In the user perception, an operation comes together with the page where it can be invoked. At design level, it must be associated with the conceptual object that defines the page. The problem may occur when a conceptual model supports abstractions both for simple "objects" directly materialized into individual

---

[2] A complete design would require that to specify also not only the types of collected elements but also the criterion to select them, the collection internal organization and its navigation features.

"pages", and for composite "objects" corresponding to several pages: Should operations be modeled as attributes of simple objects only, or also of composite objects? What is their semantics in the latter case? We believe that in a number of situations the designer may need to attach operations to both simple and composite objects, and we suggest that, in the case of composite objects, the operation *scope* includes to all object constituents. Let us consider the following examples. A virtual museum application may have a *create tour* facility that allows users to collect the art works of their interest and to create their own collection of pieces across which they can navigate linearly. If an art work is composed of several components (pages), each one discussing a different aspect of the art work, the designer may offer the possibility of including in the tour either a specific page only, or all pages that represent the art work in one shot. In the first case, the *include tour* operation should be assigned to the simple conceptual object (e.g., "introduction to the art work") corresponding the specific page, and, in the second case, to the composite object "art work" corresponding to all art work pages. In an e-commerce application, a useful operation is *convert to currency X* that allows users to visualize the cost of a product using the currency they prefer. This could usefully be invoked from any product page, but also from any set of products (this would display the price, according to the chosen currency, for all products in the set). To model this situation, the *convert to currency X* operation should be associated both to the simple objects corresponding to products and to the composite object representing the collection of products.

Finally, web applications may require that operations be attached to navigation structures, i.e., links. Consider, for example, a fashion web application in which different product lines are described and each line comprises a number of different products (shirts, pants, dresses, etc.). When users traverse the link from a product line to related products, they may need to filter them according to different criteria, e.g., size, color, use. In this case, the filter operation should be modeled as a property of the link rather than of the product line.

4. Evolution in Web Applications

Differently from traditional web sites, hypermedia structures of web applications are very often dynamic entities, in the sense that information and navigation objects evolve either along the time, or by (direct or indirect) effect of user operations.

Consider, for example, a conference manager system, i.e., a web application that must advertise a scientific conference, enable authors to submit papers (in two steps: abstracts and full papers), support program committee members in reviewing and selecting papers, and help the chair setting up the conference program. An obvious entity type in this application is *Paper*. When an author executes the operation *submit abstract* (from the *how to submit* entity), an instance of *Paper* is created, but only its *Abstract* component exists. Once the deadline for submission expires, the hyperbase must evolve: each paper has a new (optional)[3] component of type *Submission*, which is cre-

---

[3] The *Submission* component is optional since authors may submit abstracts and then renounce to submit full papers.

ated when authors invoke the *submit full paper* operation. When the review process is over, i.e., the program committee has finally executed all needed ranking operations, the structure of the paper has another evolution step: A new component of type *Camera Ready* is added and it replaces the previous *Submission*[4].

Evolution may involve also collections. The collection *all reviewed papers*, for example, contains all papers in state *reviewed*. Thus new elements are continuously added as long as all papers are reviewed. Moreover, it is replaced by the collections *definitively rejected papers* and *definitively accepted papers* only after the program committee has finally discussed all papers, which are either *rejected* or *accepted*.

Similarly, also semantic associations evolve: the semantic association of type *presented-in*, which connect each accepted paper to the session in which it is presented, exist only for *accepted* papers and after that they are accepted.

Finally, evolution defines also whether an operation can be invoked or not (and sometime the way the operation is executed, as discussed in the following section). For example, the operation *include paper*, which is available from each session entity and allows the program chair to set what papers are presented in a session and to link the session page to its papers, becomes available only when the review process is over and applies to *accepted* papers only.

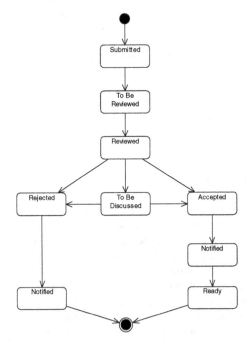

Figure 2: Statechart Diagram for the *Paper* entity type

---

[4] We assume that the *Abstract* component always remains available.

Modeling the dynamics of hypermedia structures is a novel task with respect to traditional web design, and can be expressed with various means. W2000 models the states along which the various types of hypermedia elements (entities, associations, and collections, in the HDM jargon) can evolve by borrowing *statechart diagrams* from UML. The statechart diagram of Figure 2 models the evolution of *Paper* entities.

After introducing evolution and states, we can generalize how designers should exploit statechart diagrams:

- They must specify a statechart diagram for each "relevant" hypermedia element, that is, for each element that can evolve;
- They must decide what states are perceivable by users (and therefore must be explicitly modeled also as information elements in the hypermedia schema);
- They must enrich the hypermedia schema with formal or informal annotations (comments, cardinalities, or OCL[5] constraints) that correlate all possible states of an element with possible variants of its structure and content. An example is shown in Figure 3: It describes the structure (in-the-large, omitting all details about attributes) of the *Paper* entity type, pinpointing that different components exist in different states of an element of this type.

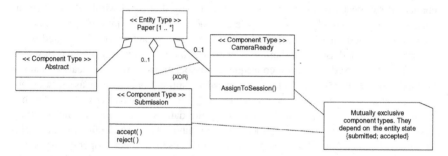

Figure 3: The Structure in-the-large of Entity Type *Paper*[6]

5. Operation Dynamics

Designing web operations is slightly different from designing conventional operations, for a number of reasons:

- Web operations do not operate on generic data structures, but on peculiar structures that capture hypermedia specific abstractions. As such, they can be designed on top of a hypermedia schema.

---

[5] OCL stands for Object Constraint Language and is the textual notation offered by UML to annotate graphical elements.

[6] Designers could exploit also inheritance to make similarities among elements explicit. Containment characterizes the structure of composed elements.

- Most operations for web applications can be invoked during navigation (and may need to take into account the navigation context as an implicit parameter), and they may require some navigation steps to be completed.

These peculiarities impose designers to adapt traditional techniques for modeling operation dynamics to address the characteristics of web operations and to exploit built-in hypermedia concepts.

Traditional object-oriented modeling specifies an operation as a set of cooperating objects that exchange information and require services each other. Every involved object knows directly, through references or pointers, all objects it can interact with. Only the notion of *object* (with data and functional attributes) is built-in in the model; everything else must be designed explicitly.

In contrast, web operations work on a variety of "objects" of different complexity, as we discussed in Section 3: simple, elementary objects (e.g., components, in HDM, corresponding to single pages) very much similar to conventional objects, and clusters of objects, i.e., composite objects (e.g., entities and collections, in HDM) corresponding to several pages. Thus, compared to traditional object-oriented design, the data model exploited by web operations should deal explicitly with these hypermedia structures that have different levels of granularity. In web applications, objects are connected by means of associations ("links") among them and within their constituents. Associations should not be considered (and designed) as static properties of objects. Links "outgoing" from a hypermedia object are not automatically created when the object is created. Some of them may even depend on the run time context in which the object is accessed [14]. As a consequence, associations in web applications cannot be modeled either as simple references (pointers) among objects, nor as parts of objects. Treating links as first class elements has been traditionally accepted by the hypermedia community, sometimes accepted by the database community, and not fully acknowledged by the software engineering community. For web applications, associations *must* become first class citizens in any model used to specify operations: operations can create, modify, and delete links explicitly (or implicitly) as if they were objects and can be included as link attributes.

In traditional object oriented design, any operation is designed through explicit methods (and possibly special-purpose objects) and any interaction among objects through method invocation and message passing. Only methods for creating and deleting objects are built-in. Navigation, if needed, must explicitly be designed. In web applications, which intensively involve navigation, the designer should not be forced to redefine each time the *follow a link* operation: Basic navigation, with the user moving from one piece of information to another, should be a built-in mechanism, complementary to method invocation. The designer, for example, may need to specify that an operation can be invoked from an object of a given type (entity, collection, or association), no matter how the user has reached that object during a session. From the above consideration it follows that the modeling notation must support explicitly a specialized mechanism of message passing: *free navigation*, that is, the possibility of showing only the final destination of some navigation steps without specifying the followed path. Free navigation is not strictly necessary, but helps supplying a usable specification means. It is a shortcut that makes the process of identifying fixed points that users must reach to complete the execution of a given operation or to invoke it lighter and simpler.

An opposite situation is when an operation can be invoked only in a given navigational context, or it requires, during its execution, that the user reaches a given object, but only under some constraints. Consider, for example, the operation *include paper* that can be invoked in a web-based conference manager from an entity of type *Session*. Clearly, *include paper* requires that the paper to be included in a session be identified, and is among *accepted* ones. A possibility (among several others) is to require that the user navigate to the collection *definitively accepted papers*, and chooses the paper from this context. The above design choice can be described by means of another specialization of message passing: *constrained navigation*, that allows the designer to specify a specific path as a precondition for invoking a given operation or for completing it.

All peculiar requirements of modeling web operations described so far can easily be met by extending UML *interaction diagrams*. For example, Figure 4 presents an extended UML sequence diagram that shows what a conference chair should do to create a new conference session and associate a paper with it in a hypothetical conference manager. The following are the main required extensions:

– We organize objects in different categories: entities, associations, and collections, rendered with different shapes. Different object types also imply different rules for message passing, but we have no space to discuss them here.

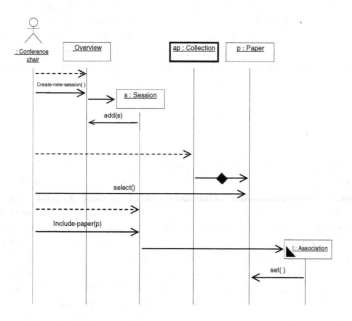

Figure 4: An Example of Extended UML Sequence Diagram

- We refine message passing with constrained and free navigation. Constrained navigation is depicted using a line with a diamond, while free navigation is rendered with a dashed line.

The simple example of Figure 4 specifies that the conference chair can freely navigate to the conference *Overview* entity and ask for creating a new session. This user operation makes the system create a new empty component - session *s* - and adds it to entity *Overview*. Then the chair navigates "freely" to the collection of accepted papers (*ap*) and follows a specific association (see constrained navigation arrow) to identify and select a particular paper *p* in the collection. After browsing back to session *s*, (s)he invokes the *includes-paper* operation to add *p* to the session. The user operation creates a new association (*l*) and links *s* to the selected paper *p*.

A final concern is on the detail level that should be achieved. As discussed in Section 3, operations can be organized in: *user operations* and *system operations*. User operations are the only operations directly visible to users, thus callable by external actors. System operations are not visible to users, but are necessary to and triggered by user operations. Among system operations, we can distinguish between *base operations* and *application-specific operations*. The first set comprises all general-purpose functions that operate on hypermedia structures (e.g., to add an entity to a collection, or to add a component to an entity): they can be considered as built-in in any conceptual design model and left unspecified during the design process. The second set comprises system operations that are application specific and must be defined by designers (e.g., the operation to compute V.A.T. for a product once the product has been definitively bought). In conceptual modeling, it is important to describe the application from a user-oriented perspective, i.e., designing information/navigation structures and operations that are to be manipulated by end users rather then by an implementation environment. Therefore the detailed description of system operations can be considered as part of implementation design, and the primitives for their specification can be omitted in a conceptual model for web applications.

6. Conclusions

This paper discusses some modeling problems raised by the peculiar features of the new generation of web applications, mostly induced by the need of capturing the complex intertwining between navigation and operations of different kinds. The paper proposes also some modeling concepts that could be introduced in current models – for object oriented, hypermedia, or web sites design – and sketches how they can be rendered with suitable extensions to standard UML. The concepts and primitives described in this paper are part of W2000. W2000 merges HDM2000 and UML and provides an UML-based unified modeling framework to support the various activities involved in the design process of web applications. W2000 (not completed yet) defines also modeling guidelines and heuristics to guide the designer, and considers requirements analysis and the correlation between requirements and design choices.

The paper discusses some relevant topics, but leaves others, like consistency, customization and transaction design, untouched.

Consistency becomes a key concept when we design several different interdependent aspects using a variety of concepts and primitives. Consistency among the design specifications of different features must be carefully design, and needs to be supported by proper design tools. The advantage of adopting UML as basic notation is that UML-based CASE tools are easily available, and can support the specification activity and the needed consistency checks. For such a reason, in parallel to the conceptual definition of W2000, we are developing a set of tools to support the design in all its different aspects. This set of tools represents an evolution of the previous set of Hypermedia tools, JWeb [4], based on the previous version of HDM.

Customization must be considered when web applications may address a variety of different users with different navigational and functional needs, constraints, or situations of use. A generic user of a conference web application, for example, is not allowed to see the reviews of papers; authors could be allowed to access only a selection of the reviews, i.e., those concerning their papers; in addition, an author could not be entitled to use the link connecting a review with the reviewer. At a lower design level, specific information items could become not visible to certain roles under certain circumstances.

Web application customization is a complex task since several facets must be considered. Beside user profile and role, customization choices may depend (sometime simultaneously) on different factors:
- The overall state and evolution of the web application (e.g., the single reviews may be invisible to authors until all reviews have been completed)
- The situation of use. Web applications are becoming multi-device and ubiquitous, and are (or will become soon) accessible everywhere through a number of different devices, ranging from high-end workstations, to PC's, to PDA's, to cellular phones, etc. The "same" application needs to be customized for each device, and to take dynamically into account the physical situation in which it is used.

To provide a global framework where customization could be addressed, we are working on refining the notion of *hyperview*, as a conceptual mechanism in W2000 capable of supporting advanced customization.

Transaction design is need to define the execution semantics of operations, at a lower abstraction level. The two main problems, still to be touched by our research, are how to implement *atomic transactions* in an inherently volatile and unstable environment as a web application and how to define *long-lived web transactions*. The second aspect is of particular relevance for web application: A browsing/navigation session may take minutes, hours or days. Within a session a user may perform a number of operations (e.g., reservations in a tourism application) that at the end must collapse in a single transaction, in the sense that all of them must be committed. These issues have been solved only partially so far and are part of our research plans.

References

1. P. Atzeni, G. Mecca, and P. Merialdo: "Design and Maintenance of Data-Intensive Web Sites", Proceedings of EDBT 1998, pp. 436-450.
2. L. Baresi, F. Garzotto, P. Paolini, and S. Valenti: "HDM2000: The HDM Hypertext Design Model Revisited", Tech. Report, Politecnico di Milano, Jan. 2000.
3. H. Baumeister, N. Koch, and L. Mandel: "Towards a UML extension for hypermedia design", in Proceedings of UML´99, LNCS 1723, Fort Collins,USA, October 1999, Springer Verlag.
4. M. Bocchichio, R. Paiano, and P. Paolini: "JWEB, An HDM Environment for Fast Development of Web Applications", in Proceedings of the IEEE Multimedia Computing and Systems, 1999.
5. G. Booch, J. Rumbaugh, and I. Jacobson: "The Unified Modeling Language User Guide", Addison Wesley, 1998.
6. S. Ceri, P. Fraternali, A. Bongio: "Web Modeling Language (WebML): A Modeling Language for Designing Web Sites", to appear in Proc. Of Int. Conf. WWW9, Amsterdam, May 5 2000.
7. P. P. Chen, D. W. Embley, and S. W. Liddle eds, Advances in Conceptual Modeling - Proc. WWWCM99 -Int. Workshop on the World Wide Web and Conceptual Modeling, (Paris, Nov. 1999) Springer-Verlag, LNCS 1727.
8. J. Conallen: "Building Web Applications with UML", Addison-Wesley, 2000.
9. J. Conallen: "Modeling Web Application Architectures with UML", Communications of the ACM, 42:10, Oct. 1999, pp. 63-70.
10. L. Delcambre, D. Maier, "Models for Superimposed Information", In [5].
11. O.M.F. De Troyer, C.J. Leune, "WSDM: A User Centered Design Method for Web Site", in Proc. of Int. Conf. WWW7.
12. P. Fraternali, P. Paolini: "A Conceptual Model and a Tool Environment for Developing More Scalable, Dynamic, and Customizable Web Applications", Proc. EDBT 1998, pp. 421-435.
13. F. Garzotto, P. Paolini, D. Schwabe, "HDM - A Model-Based Approach to Hypertext Application Design", TOIS 11(1) (1993), pp.1-26.
14. F. Garzotto, L. Mainetti, P. Paolini, "Navigation in Hypermedia Applications: Modeling and Semantics", in Journal of Organizational Computing and Electronic Commerce, 6 (3), 1996.
15. T. Isakowitz, E. Stohr, P. Balasubramanian: "RMM: A Methodology for Structured Hypermedia Design", CACM (1995), 38(8), pp. 34-44.
16. G. Rossi, D. Schwabe, F. Lyardet "Web Application Models are More Than Conceptual Models", in [5].
17. D. Schwabe, G. Rossi, "An Object Oriented Approach to Web-Based Application Design", Theory and Practice of Object Systems, 4 (4), J. Wiley, 1998.

Using Webspaces to Model Document Collections on the Web

Roelof Van Zwol and Peter M.G. Apers

University of Twente, Department of Computer Science
P.O.box 217, 7500 AE, Enschede, the Netherlands
{zwol, apers}@cs.utwente.nl

Abstract. Due to the unstructured character of data on the web it is hard to find specific information when surfing over the web. Search engines can only rely their results on IR techniques available, and most of the time they lack the desired power in query formulation. Modelling data on the web, as if it was designed for use within databases, provides us with the necessary basis for enhancing the query formulation. This requires special care for dealing with the included multimedia data and the semi-structured aspects of the data on the web. Modelling the entire web would be too ambitious, therefore we focus on a more feasible environment, like the intranet, where one can find large collections of related data. This article describes the webspace method for modelling the content of a collection of a domain specific documents, and offers a solution for the above mentioned problems.

1 Introduction

Surfing over the WWW, one is likely to find large collections of electronic data, made available via the web. Most of the time, despite all standardization efforts, the format of this data does not apply to a rigid structure. Therefore it cannot be stored easily, using existing relational or object-oriented models [11,3]. When modelling web data, the focus used to be more on the presentation of data, than on the content (modelling) aspect. Besides the structure of data, various types of multimedia are also involved, hardening the task of modelling data on the web.

The contribution of this paper is twofold. First it describes our approach for conceptual modelling of large collections of related data, using webspaces. Secondly, it discusses the webspace modelling tool, set up to guide the process of modelling documents for a webspace. In this tool, modelling documents for a webspace is split up in four tasks, guarding the design process of a webspace.

A webspace consists of two levels. At the document level a webspace can best be seen as a collection of documents describing related information. The documents can contain all kinds of multimedia data and have an irregular structure, as long as the content of the documents is related to a specific domain. At the semantical level of a webspace, concepts are defined, describing the content of the documents at a semantical level of abstraction. These concepts are modelled

S.W. Liddle, H.C. Mayr, B. Thalheim (Eds.): ER 2000 Workshop, LNCS 1921, pp. 101–114, 2000.

in a webspace schema, using existing object-oriented modelling techniques.

Based on the model for webspaces, modelling aspects for data put on the web are studied. Since data on the web is considered to have an irregular structure, modelling such data is hard. Following an object-oriented approach, the same 'web objects' do not necessarily have all the attributes present. The same attribute might also have different types in different objects. Even worse is when semantically related information is represented differently in different objects. Data that inhabit these characteristics is called semi-structured data [15]. To deal with this efficiently three types of semi-structuredness have been distinguished in our webspace model. To be able to deal with the various types of multimedia involved, our concept-based model for a webspace is combined with a content-based framework for information retrieval.

Conceptual modelling forms the key issue when defining new documents for a webspace. This process is guarded by the webspace modelling tool. There an author starts with identifying concepts, needed for the document. These concepts are modelled in a webspace schema, as if a new database schema is designed. The next step is to define the structure of the document to be created, by specifying a view over (a part of) the webspace schema. Once the structure of the document is known, content can be added and finally a presentation for the document has to be defined. The latter states the contrast with the way web-documents are designed now-a-days, where one starts with specifying the layout, before adding the content and where modelling aspects are considered to be a minor issue.

Once a webspace is defined properly, at both the document level and the semantical level, the webspace schema can be used to generate a schema for our object server, using a meta-database approach. Based on a webspace schema, a webspace consisting of a collection of related documents, can be searched by formulating queries over the webspace schema, using the meta-data stored in the object server. Instead of querying single documents, one can formulate a query over information stored in separate documents [18].

State of the Art Modelling data on the web is an ongoing research area, where many research projects have been positioned around. Closely related to our approach is the Araneus project [10] where also existing database technology is applied to the WWW. The web documents and hyper texts are modelled and stored using a relational data model, called ADM. This project is also concerned in handling both structured and semi-structured documents. The main difference with our approach is that we aim at combining concept-based search with content-based information retrieval, to come up with new query formulation techniques for data on the web. In [4], [13], and [7] a relational approach is used to store semi-structured data efficiently. The difference is that we use the Moa object algebra on top of the vertically fragmented data model and that the semi-structured data itself is not stored in the database, but only the meta-data that can be derived.

Others, like in Lore [9], XML-QL [5] use the structure of the XML document as their model. Adding support for regular path expressions does enable them to search for patterns and structure in the XML data. Except for [8], where text-based search is integrated into XML-QL, content-based queries are not supported over complex multimedia documents.

Organization of this Paper The paper is organized as follows. Section 2 discusses the webspace method. There we focus on the webspace model, and discuss the semi-structured- and multimedia aspects. Finally an overview of the complete webspace architecture is given. In Section 3 the webspace modelling tool is discussed. The process is illustrated by an example Section 4. Finally we will come to our conclusions in Section 5.

2   Webspaces

Searching for data on the web can become more powerful, when this data is modelled using database techniques. We already argued that this cannot be done straight forwardly, since the web is not a database [12]. When focusing on a more limited environment, like defined for a webspace, the techniques developed for databases can be applied. By following this approach two major obstacles were encountered. Dealing with data on the web, also implies dealing with multimedia. Others, like [17] already argued for the need to extend database technology to deal with these types efficiently in a database environment. As will be explained in Section 2.3 we adopt the Mirror framework for content and multimedia DBMS [16] for dealing with such data efficiently. The second problem that has been dealt with is the semi-structuredness of the data involved. The current trend for dealing with such data is by adopting a graph-based data model, most of them are based on the Object Exchange model [9]. We have chosen not to follow that approach, but instead an object-oriented approach is used to deal with such data efficiently (Section 2.2). In our model for a webspace these two research areas are brought together, to come up with better modelling facilities for data on the web and more powerful query mechanisms for large collections of related data.

2.1   The Webspace Model

When looking at a collection of related documents, it is possible to identify a (finite) set of concepts, which describe the content of the documents at a semantical level [18,1]. In our model for webspaces this is exploited by identifying two levels. At the document level one can find a collection of related documents, which should be (made) available to the WWW. At the semantical level the before mentioned concepts should be defined and modelled in a object-oriented schema, called the webspace schema. Section 2.1 describes how webspaces should be defined at the semantical level in a webspace schema, and in Section 2.1 it is explained how this relates to a webspace at the document level.

Semantical Level For each webspace a set of concepts should be defined, describing the content of the documents involved. Such concepts are identified by a unique name, and should have a semantically well-defined definition. The semantical level is then formed by a webspace schema. This schema is based on an object-oriented data model, which allows concepts to be modelled in terms of (1) classes, (2) attributes of classes, and (3) associations over classes. The model also includes a generalization mechanism, allowing classes to be defined as subclasses of other classes. Together, the set of classes, attributes, and associations form a partition which is equal to the set of concepts defined for a webspace.

Once a concept is defined to be a class in the webspace schema, it cannot be reused as an attribute or an association in the same webspace. Likewise for the attributes and associations involved. Since authors publishing their documents on a webspace cannot be assumed to have the skills of a database administrator, notion of types, and cardinality is left out of the data model at this high conceptual level of abstraction. To our opinion such problems should not be visible to the users, and should be dealt with at the logical and physical level of the webspace system, as is discussed in Section 2.4.

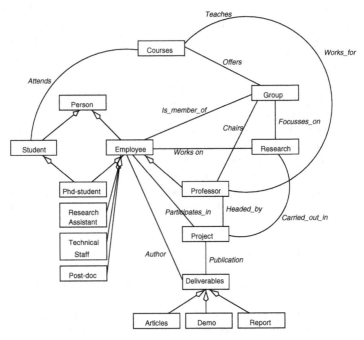

Fig. 1. Webspace Schema for the Homepages of the Db Group

In Figure 1 an example of a webspace schema is given, based on the webspace that is set up for the home pages of the database group. The fragment shown here describes the webspace at the semantical level based on the concepts needed for the general database home pages. All the concepts shown in the figure are modelled either as classes, or as association over classes. To keep the figure viewable, the attribute concepts, defining the properties of the class concepts are left out in this figure. In Section 4 this example is worked out in more detail.

Document Level The document level of a webspace stores the data involved. For this purpose XML is used to mark up the data. When the data involved has a rather structured character, it is very well feasible to store the data using a regular DBMS, as proposed in [11,6] and others. Although it is also feasible to deal efficiently with more semi-structured data stored in a DBMS [7,19,13], we have chosen to store the data in XML-documents. In the next section the semi-structured aspects of the webspace data model are discussed briefly, showing how the problem is dealt with at a meta-data level
With XML one can easily mark up the content of a document, thus allowing authors to make the structure of their documents explicit. Others use this tree-based structure directly as the input for searching through the content of single documents. But only basing the search on the structure might lead to semantic misinterpretations. To bridge this semantical gap with the user searching the data within a webspace, concepts are defined in a webspace schema providing a semantical layer over the collection of documents involved. From that perspective, each element and attribute used in an XML-document, should correspond to a concept defined in the webspace schema.
Following this approach, ensures that any document can be seen as a materialized view of the webspace schema. In Table 1 a fragment of an XML document describing a project is shown. The root element of this document corresponds to a class concept of the webspace schema of Figure 1. The document describes a 'project' object, having some of the attribute values, specified in the webspace schema available. It also describes relationships with other web objects, like an article, report, demo, and an employee. Figure 2 shows the part of the webspace schema that is materialized in this document fragment.

Fig. 2. Fragment of Webspace Schema

2.2   Semi-structured Aspects of the Webspace Data Model

Due to the semi-structuredness of the data on the web, the data model for modelling the webspace schema also differentiates from other object-oriented

Table 1. XML Document for Project of the Db Group

XML-document: mqp-project
$<?xml\ version =' 1.0'\ encoding =' UTF - 8'? >$
$<!doctype\ project\ system\ 'project.dtd' >$
$< project\ title =' Multimedia\ Query\ Processing'$
$\qquad startdate =' 1\ may\ 1998'$
$\qquad enddate =' 1\ may\ 2002' >$
$\quad < description >\ ...\ < /description >$
$\quad < publication >$
$\qquad < article\ title =' Modelling\ ...\ Moa'\ year =' 1999' >$
$\qquad\quad < author\ name =' ...' >\ ...\ \ ...\ $
$\qquad < /article >\ ...$
$\qquad < demo\ ...\ / >$
$\quad < /publication >\ ...$
$< project >$

data models. For our model this is expressed by identifying 3 levels of semi-structuredness.

- **Type level**. In a regular data model each attribute will have only one type assigned. In the webspace data model each attribute is allowed to have more than one type. Thus, a date attribute can be specified as the string '20 April 2000', but in a different document it can be defined as the integer '200400'.
- **Attribute level**. Objects of the same class do not necessarily have the same attributes available, as long as the webspace schema contains all the attributes used for that object's class. Although not obliged, it is possible to define a unique key for some attributes of a class. When constructing the meta-database, undesired duplications of web objects can then be avoided.
- **Class level**. Here authors can specify synonyms for the concepts defined at the semantical level. These synonyms should form a set of alternate concept names, each related to only one of the concepts already in use for the basic webspace. For example, one might want to introduce the synonym concept 'automobile' for the concept 'car'.

The first two aspects are dealt with in Moa at the logical layer of the architecture given in Section 2.4. In [19] is described how is dealt with problems like type coercion and semi-structured data modelling.

2.3   Dealing with Multimedia Information

Data on the web more and more inhabits a rich multimedia character. When querying the content of a webspace, we can no longer simply fulfill with text-based search. It is also desired that other types of multimedia can be included during the search process. Therefore we have extended our framework for concept-based search with content-based information retrieval. For this purpose we have adopted the Mirror framework for content-based multimedia databases

[16] and used the multimedia extensions available for the object server (see Figure 3).

Extracting the meta-data from the documents, is done by using a daemon architecture. A daemon in this context can best be seen as a black box, which given multimedia object, like a text fragment or an image, extracts the meta-data from that object. This data can then be stored in the underlying meta-database. The daemon architecture makes use of a daemon data dictionary, where the daemons are administrated. When defining new concepts for a webspace, the author needs to specify how a daemon is invoked on a web object, by specifying a trigger for that daemon. A trigger consists of three parts, which contains information about which classes, attributes and types of the webspace a daemon should work. For example the following string representation of a trigger: ' * . abstract . * ', if assigned to the text daemon, would startup the text indexing process, whenever the attribute abstract is found in one of the documents available on the webspace. The wildcard notation is used to indicate that the other parts of the trigger string do not matter, in this case it concerns the class- and type part.

2.4   Architecture of the Webspace System

To implement the ideas for modelling and querying the content of webspaces the architecture as shown in Figure 4 is set up. It is based on a three layer architecture, consisting of a physical, logical, and conceptual layer. The work described in the second part of this article mainly focuses on the top left side of the conceptual layer. Going bottom up through the figure, the following parts are identified.

- **Object server.** The object server is formed by the physical layer and logical layer of the architecture. At the physical layer the meta-data obtained from the XML documents is stored in either Monet [2], a binary relational database kernel, or Postgres [14], a object relational DBMS. On top of both databases we use Moa [2,16]. Moa consists of a structural object-oriented data model and algebra. This provides us the desired physical data independence at the logical layer. Moa uses an extensibility mechanism, which on one hand adopts the extensibility mechanism provided by the physical layer, but also provides structural extensibility, which is used to model the semi-structured aspects of the webspace model at both the type, and attribute level efficiently [19,7]. The structural extensibility is also used to implement the framework for content-based multimedia information retrieval as proposed by [16].
- **Web object retrieval and storage layer.** One of the central components in the webspace system is the layer responsible for retrieving the meta-data from the documents available. Using the webspace schema supplied by the authors of the webspace, several intermediate schemas are generated to populate the object server. As output this component delivers a Moa schema at the logical level, and a physical schema used by Monet (MIL), and Postgres (PSQL). Secondly it is responsible for obtaining web objects from the

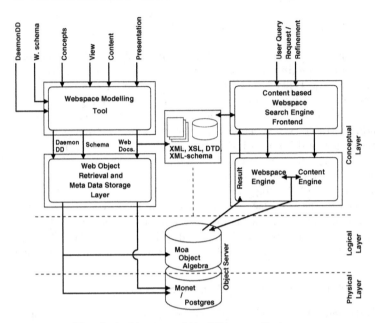

Fig. 3. Architecture of the Webspace System

XML documents and retrieving all kinds of meta-data from the various types
of multimedia involved. For this purpose we use a daemon data dictionary,
which is triggered when a web object belongs to a certain class, contains
some attributes or if the attribute is of a certain type. For example, for each
object belonging to the class 'Image', and having the attribute 'src' a daemon will be assigned to extract RGB and HSB information from that image.
The same is done for text fragments. In that case a text indexer is started.
When all the meta-data from one web object is extracted, it is stored at the
physical layer, using either Monet, or Postgres.

- **Webspace modelling tool.** In the next section we will discuss this component in detail. It is responsible for all the tasks that need to be performed
  in order to set up a single webspace, at both the semantical level, and the
  document level.
- **Webspace and content query engine.** This component consists of two
  'engines' one for composing queries regarding concept-based search on a webspace, and one for doing content-based querying over the multimedia involved. Work still needs to be done here, in order to realize the full integration
  of concept-based search and content-based information retrieval.
- **Webspace query front end.** With this front end we intend to offer the
  user a query interface that allows him to compose complex queries using a
  graphical notation of the webspace schema and combine this with content-
  based retrieval techniques.

3   Webspace Modelling

This section describes how to model concepts defined for a webspace, and how new documents can be created for a webspace, by defining a view over the webspace schema. The process of modelling a webspace is split up in four tasks, which are carried out sequentially. In practice it is desirable to go back one or more modelling steps (tasks) and to be able to add new concepts and change a documents view, while adding the content to the documents structure. Figure 4 shows the architecture of the webspace modelling tool in detail. From the user point of view it offers an integrated environment in which all the modelling tasks are performed. It consists of four components, each referring to one of the tasks that have to be performed.

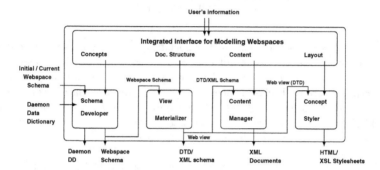

Fig. 4. Webspace Modelling Tool

Each of the four modelling tasks have to be carried out when defining new documents for a webspace. The first task deals with modelling the webspace schema and is discussed in Section 3.1. During the second task the author has to specify which concepts should be used for his document, by creating a view on the webspace schema (Section 3.2). Once the structure of the document is known, content should be added by the author for all the required fields. Table 2.a shows the in- and output requirements for the XML content tool. For this purpose we integrated one of the currently available XML-tools in our webspace modelling tool. With this XML-tool XML documents can be created, when given a DTD (Document Type Definition). During the last task the author needs to specify a presentation for the XML document. Table 2.b shows the in- and output requirements for this task. Again we use an already existing tool for this purpose. The XML styling tool generates an XSL (Extensible Stylesheet Language) style sheet used for the presentation of the materialized view on the webspace defined in an XML-document.

3.1   Webspace Schema Modelling

During the first task of the webspace modelling process, the author should take care of modelling concepts at the semantic level in a webspace schema. The

Table 2. XML content and presentation tools

input	output
• DTD / XML schema • content information	• A materialized view on the webspace, in the form of XML documents.

(a) XML content tool.

input	output
• Web view. (in the form of a DTD) • Layout information.	• HTML/XSL Style sheets.

(b) XML styling tool

schema modelling tool lets the author specify what concepts should be defined in terms of classes, attributes of classes, and associations over classes. Since all concepts are related, the resulting webspace schema can be presented as one graph, where each class concept is related by an association concept to at least one other class concept. All the attribute concepts defined for a webspace should also be connected to a class, or association concept. The following steps are carried out during the schema modelling task.

- Identification of concepts needed for the document to be defined.
- Classifying them as class-, attribute-, association, or synonym concepts.
- Defining the properties of a class concept by adding attribute concepts.
- Connecting class concepts through the association concepts.
- Assigning generalization connections between classes.
- Specifying for each attribute what types that attribute can have.
- Specifying daemons that can be invoked on an web object, by specifying triggers.

Table 3. Webspace schema modelling

input	output
• initial webspace schema. • daemon data dictionary. • concept information.	• webspace schema (altered). • daemon data dictionary(altered).

In Table 3 the in- and output requirements for this tool are specified. Most of the time one will start with an initial schema and will only have to perform some of the actions mentioned above. Notice that the author can assign several types for a single attribute and that the author can decide which daemons should be invoked on a web object by defining a trigger for that object. Of course when the author needs to modify parts of the schema, a new webspace schema will be generated, but the author is not allowed to change the schema in such a way that a conflict can exists with the webspace at the document level. For example if in one of the documents a concept is used, it cannot be deleted from the webspace schema. When new triggers are assigned to a daemon, the daemon data dictionary also has to be updated. Defining new types and adding new daemons is typically a task done by the administrator of the webspace. Since this will influence the underlying layers of the webspace system.

3.2 Webspace Views

Once the semantical level of a webspace is up to date, views on the schema should be specified, to create new documents at the document level of a webspace. First an object oriented view on the webspace schema is created. During the following steps, this view is transformed into a tree presentation, which finally results in a DTD specification of the new document. The following steps are carried out when specifying views on the webspace schema.

- Select all concepts that will be used in the document, the result should be a connected graph.
- Select the class, which will form the root element of the XML-document.
- Specify for each class concept, which attributes will be used in the docuemnt.
- Specify whether a class attribute will be used as an attribute or as an element in the XML-document.
- Finally specify the content model for all the elements of the DTD, i.e. the occurrence indicators (? * +) and connectors (, |) need to be specified

As the result of this task we have chosen to create the DTD of a document. Alternatively it is also possible to generate XML schema specifications, instead of the DTD of a document.

Table 4. view tool

input	output
• Webspace schema. • Web view information, in terms of concepts.	• web view, in the form of a DTD or XML schema.

4   Webspace Example: The DB-Home Pages

In this section we will illustrate the webspace modelling process by some examples. Everytime the webspace chances at the semantical level, we need to update the webspace schema and the daemon data dictionary. Both are exported to the lower levels, using XML as an exchange format.

With this example we will show how the structure of an XML-document is created, by defining a view on the webspace schema. We will base our example on the project document, as presented in Table 1. One of the first steps, when defining a new view, is selecting all the relevant classes, associations and attributes. For our view, we have chosen to select the classes: 'Project', 'Article', 'Demo', and 'Report'. Each view on a webspace schema needs to be a connected graph, thus the associations: 'publication', and 'author' are also selected. Finally we have to choose what attributes will participate in the view. Once this is done the view is gained, as shown in Figure 2. The next step is assigning which class will form the root element in the view. Once this is known, the view can be seen as a tree (Figure 5.a). To complete the transformation of the view into a DTD, we still have to decided which attributes of classes will appear either as elements or attributes in the final DTD. Finally the connectors and occurance indicators need to be specified for all the elements and attributes of the DTD. In Figure 5.b the transformation to the documents structure is completed.

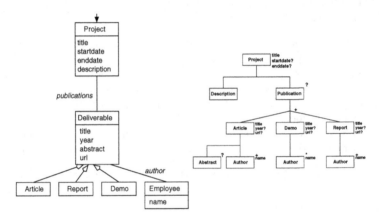

(a) Tree of project document   (b) DTD for project document

Fig. 5. tree and dtd representation of the project page

For the graphical representation of the DTD we used the notation of the XML dtd specification. Notice that the superclass 'Deliverable' has disappeared from the view during the last transformation, since this class was never selected in the first place. Furthermore the association 'author' uses the primary key of 'employee', to specify the relation of an article with its author, which apparently turns out to be an employee. The elements 'Abstract' and 'Description' of Figure 5.b were originally modelled as attributes in the webspace schema, but in the view these attributes are defined as child elements referring to the same class. Once the content is added to the structure and a presentation is defined, using XSL-T, the document is ready to be shown.

5   Conclusions

When surfing over the WWW one is likely to find large collections of documents, containing information regarding a specific domain. Unfortunately searching such collections of data still is not easy, since powerful search engines for querying such collections are still missing. Modelling these collections of data, based on existing ideas for modelling data using database technology, will bring the power of query formulation as currently available for databases to the web. With the webspace method we model such collections of data, by introducing a semantical level of abstraction. At this semantical level concepts used at the document level are modelled in a schema for a webspace, as if one would design an object-oriented database schema. When the concepts needed are defined at the semantical level, views on the webspace schema can be defined,resulting in XML documents at the document level of the webspace. In the light of this model two aspects have been discussed. The first aspect concerns the semi-structured aspect of the data involved. The webspace model is capable of dealing with such problems efficiently. The second aspect regards the various types of multimedia that are used in the documents. Integrating our approach with the content-based framework for multimedia databases tackles this problem.

In the second part of the article the webspace modelling tool is discussed, which guides the process of setting up and maintaining webspaces properly. This process starts with modelling concepts at the semantical level, it is shown what steps are needed to derive a view on the webspace schema and how this results in the structure of a new document at the document level. By adding content to this structure and defining a presentation for this document, the entire process of creating documents for a webspace is covered in this tool. It also supplies all the necessary ingredients for populating our meta-database used for querying a webspace.

Acknowledgements We wish to thank Arjen de Vries and Erik van het Hof for their work on content-based multimedia databases. They supplied the basis for our daemon architecture for content-based information retrieval. We also wish to thank Robert van Utteren, for his work on the implementation of the Webspace Modelling Tool and parts of the underlying layers.

References

1. M. Agosti, R. Colotti, and G. Gradenigo. A two-level hypertext retrieval model for legal data. In A. Bookstein, Y. Chiaramella, G. Salton, and V.V. Raghavan, editors, *Proceedings of the 14th Annual International ACM SIGIR Conference on Research and Development in Information Retrieval*, pages 316–325, Chicago, Illinois, October 1991. ACM Press.
2. P. A. Boncz, A. N. Wilschut, and M. L. Kersten. Flattening an Object Algebra to Provide Performance. In *Proceedings of the IEEE International Conference on Data Engineering (ICDE)*, pages 568–577, Orlando, FL, USA, February 1998.

3. V. Christophides, S. Abiteboul, S. Cluet, and M. Scholl. From structured documents to novel query facilities. In *proceedings of SIGMOD94*, 1994.

4. A. Deutsch, M. Fernandez, and D. Suciu. Storing semistructured data with STORED. In *proceedings of the ACM SIGMOD International Conference on Management of Data*, 1999.

5. A. Deutsch, M. F. Fernandez, D. Florescu, A. Levy, and D. Suciu. A query language for XML. In *proceedings of the International World Wide Web Conference (WWW)*, pages 1155–1169, 1999.

6. M. Fernandez, D. Florescu, J. Kang, A. Levy, and D. Suciu. Catching the boat with Strudel: Experiences with a web-site management system. In *proceedings of ACM SIGMOD Conference on Management of Data*, Seattle, WA, 1997.

7. D. Florescu and D. Kossmann. A performance evaluation of alternative mapping schemes for storing XML data in a relational database. Technical report, INRIA, Rocquencourt, May 1999.

8. D. Florescu, I. Manolescu, and D. Kossmann. Integrating keyword search into xml query processing. In *proceedings of the ninth international WWW Conference*, Amsterdam, the Netherlands, May 2000.

9. R. Goldman, J. McHugh, and J. Widom. From semistructured data to xml: Migrating the lore data model and query language. In *proceedings of the 2nd International Workshop on the Web and Databases (WebDB '99),*, Philadelphia, Pennsylvania, June 1999.

10. G. Mecca, P. Merialdo, and P. Atzeni. Araneus in the era of xml. *IEEE Data Engineering Bullettin, Special Issue on XML*, September 1999.

11. G. Mecca, P. Merialdo, P. Atzeni, and V. Crescenzi. The Araneus guide to website development. Technical report, Dipartimento di Informatica e Automazione, Universita' di Roma Tre, March 1999.

12. A. Mendelzon, G. Mihaila, and T. Milo. Querying the world wide web. *Journal of Digital Libraries*, pages 1(1):54–67, April 1997.

13. A.R. Schmidt, M.L. Kersten, M.A. Windhouwer, and F. Waas. Efficient relational storage and retrieval of xml documents. In *International Workshop on the Web and Databases*, Dallas TX, USA, May 2000.

14. M. Stonebraker and G. Kemnitz. The POSTGRES next generation database management system. *Commun. ACM 34*, (10):pages 78 – 92, October 1991.

15. D. Suciu. *Overview of semi structured data*, pages 28–38. Number 4. SIGACT News, December 1998.

16. A.P. de Vries. *Content and multimedia database management systems*. PhD thesis, University of Twente, Enschede, The Netherlands, December 1999.

17. A.P. de Vries and A.N. Wilschut. On the integration of ir and databases. In *Database issues in multimedia; short paper proceedings, international conference on database semantics (DS-8)*, 1999.

18. R. van Zwol and P.M.G. Apers. Searching documents on the intranet. In *proceedings of Workshop on Organizing Webspace*, Berkeley (CA), USA, August 1999.

19. R. van Zwol and P.M.G. Apers. Modelling the webspace of an intranet. In *proceeding of 1st international conference on Web Information Systems Engineering (WISE00)*, Hong Kong, June 2000.

Modeling Interactions and Navigation in Web Applications

Natacha Güell[1], Daniel Schwabe[1*], and Patricia Vilain[1, 2]

[1]Departamento de Informática, PUC-Rio.
Rua M. de São Vicente, 225. Rio de Janeiro, RJ 22453-90
{schwabe, natacha, vilain}@inf.puc-rio.br
[2]Departamento de Informática e Estatística, UFSC
Campus Universitário, Florianópolis, SC 88040-900

Abstract. This paper presents a method that bridges the gap between requirements elicitation and conceptual, interaction and navigation design for Web applications. This method is based on user scenarios, use cases, and a new graphical notation, called User Interaction Diagrams. From these specifications, it is shown how to derive a conceptual model, and then how to derive the navigational structure of a Web application that supports the set of tasks identified in the scenarios.

1. Introduction

Most current hypermedia (and web) design methods such as HDM [6], RMM [9], WSDM [3] or even earlier versions of OOHDM [13, 12] provide the designer with models and a corresponding notation to specify the design and implementation of applications. Little guidance is given as to how a designer should interact with all the stakeholders involved, capturing their requirements and eventually developing the actual design.

In this paper we present an approach to address this problem, based on a method that employs User Scenarios [1], Use Cases [10] and User Interaction Diagrams (UIDs) [15]. Whereas the first two techniques are well known in the literature, the former is a new primitive specifically designed to capture the externally observable interactions between a user and an interactive application.

The method described in the remainder of the paper (**Fig. 1**) shows how to gather requirements, how to synthesize both a conceptual design and a navigation design of an application from these requirements.

Fig. 1. Phases in the Design and Implementation of Web Applications. The shaded area is addressed by the method discussed in this paper

---

* Daniel Schwabe and Natacha Güell are partially supported by grants from CNPq; Patricia Vilain is partially supported by a grant from CAPES.

S.W. Liddle, H.C. Mayr, B. Thalheim (Eds.): ER 2000 Workshop, LNCS 1921, pp. 115-127, 2000.

The presentation in the paper will use as an example a website designed for the Training Division subsite of an intranet of a large corporation, with over 60.000 employees. This unit is responsible for offering course and training activities for the entire company, fulfilling demands from all other units, designing new courses, hiring teachers when necessary, providing the infrastructure and actually running classes.

The remainder of the paper is organized as follows. Section 2 describes how to first gather and specify requirements through user scenarios, use cases and UIDs, and how to use the collected models to derive a conceptual design for the application. Section 3 shows how to continue the design process by synthesizing the navigation design; Section 4 briefly compares the new method with existing work, and draws some conclusions.

2. Requirements Gathering and Conceptual Design

Successful designs star with suitable requirements gathered from users and other stakeholders. Requirements gathering is divided into the phases shown in **Fig. 2** identification of roles and tasks, specification of scenarios, specification of use cases, specification of user interaction diagrams, and validation of use cases and user interaction diagrams. The conceptual design presents a single phase: specification of the conceptual schema.

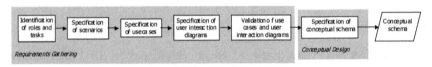

Fig. 2. Phases of the requirements gathering and conceptual design

2.1 Requirements Gathering

In order to identify the user's roles and tasks, the designer interacts with the domain to identify the roles[1] users play and tasks the application supports. This interaction is achieved through analysis of documents and user interviews.

Users may play several roles, under which they exchange information with the application. In the example, an initial examination revealed several possible roles in the design of the website: Student, Manager, Coordinator and Human Resources Coordinator. Later, during the specification of the use cases, we identified that the coordinators execute specific tasks, but the other roles had common tasks, so they were eventually merged into the single role Employee.

For each role (Coordinator and Employee), we identify the tasks the web application must support, such as (for the role Employee): Search information about a course; Search information about a teacher; Get the course material in advance.

In a second phase, each user specifies textually or verbally scenarios describing his tasks, identifying each role played if necessary. Scenarios are narrative descriptions of

---

[1]The term "role" used in this work is similar to the term "actor" used in [10].

how the application may be used [1]. Although the designers also can produce scenarios, we will concentrate on those produced by users.

In **Fig. 3**, we present some scenarios, formulated by students and managers in the case study, describing the search for a course given a subject.

AL3.4 – Finding courses given a subject

Courses can be searched by their subjects. I am a software developer. Some subjects are of interest to me, for example, "compilers" and "languages". For an administrator, subjects such as "operating system" and "tools" are interesting. So, the courses must be classified by the kinds of users.

GA5.2 – Finding information about a course

To decide if an employee can take a course, we need to get the information about the course. We would like to find the following information in the course page: program, instructor and course material. The course material is important because it can show us the organization and depth of the course.

Fig. 3. Scenarios specified by users in our case study

Next, the designer describes the use cases. A use case is a way of using the application [10], presenting the interaction between the user and the application, disregarding its internal aspects. Scenarios describing the same task are grouped into a single use case, including the information presented in all related scenarios. The designer must identify relevant data items shown in the scenarios. A use case can also be augmented with information from other use cases or by using design patterns [11].

If several types of users perform the same task, the scenarios for these types of user are grouped into a use case, identifying the roles it belongs to.

In Fig. 4, we present the use case "Browsing courses given a subject" derived from all scenarios that deal with browsing courses given a subject.

Use Case: Browsing courses given a subject.

Scenarios: ST1.4/ ST2.1/ ST3.2/ MA1.7/ MA3.6/ MA4.4/ MA 5.2/ MA7.1/ MA7.8/ C 1.1/ C3.2/ C3.3/ C4.1/ C1.12

Roles: Coordinator, Employee

Description:

1. The user enters the subject, or part of it.

2. The application returns a hierarchical set with the subjects matching the input, and the user selects the subject of interest.

3. The application returns a set with the modules of the selected subject, and the user selects one course in the module.

4. For the selected course, the application shows its name, program, weekly hours, set of subjects, set of teachers, and the set of course materials, if there are such course materials associated with it. For each teacher, the application shows the name, institution, e-mail, photo (optional) and curriculum. For each course material, the application shows the title, summary and table of contents.

5. The user can download the material (download_ftp) or see it (show).

6. If the user wishes, he can enter suggestions about the course.

Fig. 4. Use case "Browsing a course given a subject"

For each use case, we define a user interaction diagram (UID) [15]. UIDs graphically represent the interaction between the user and the application, described textually in the use cases. This diagram only describes the exchange of information between the user and the application, without considering specific user interface aspects. The UML (*Unified Modeling Language*) [2] does not include a similar diagram. UML diagrams representing interactions (sequence, collaboration, statechart, activity diagrams), consider the exchange of messages between the objects of the system. So, they are more appropriately used in the design phase, since in the requirement gathering we do not yet have knowledge about the objects. **Fig. 5** shows the UID that has been defined to represent the interactions in the task "Browsing a course given a subject".

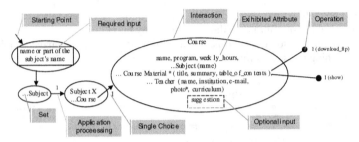

Fig. 5. User interaction diagram for the use case "Browsing courses given a subject". The meaning of each symbol type is indicated through gray callouts.

The ellipses represent the interaction between the user and the application, showing the information exchanged between them. Arrows connecting the ellipses denote that processing in the application occurs before the next information is presented. An arrow without a source represents the first interaction. A line with a black bullet in one end represents operations that do not require interaction exchange.

Once the UID has been drawn, the designer interacts with each user to validate the use cases and UIDs, obtaining their agreement. The user validates only those use cases and diagrams related with the roles he plays. Before beginning the validation with any user, the use cases have been modified according to result of the validation with the previous user, and a table is used to keep track of the successive alterations.

2.2 Conceptual Schema Specification

The conceptual design is composed of one single phase, the specification of the conceptual schema. The conceptual schema is defined according to some rules that are applied to the UIDs. Some of these rules are just an adaptation of standard schema normalization techniques [4] so that these rules can be applied to UIDs. We present some of these rules next; more details can be found in [15].

1. For each UID, define a class for each element, set element and specific element. For each defined class, assume the existence of an identifier attribute OID.
2. For each data item of the set elements that appears in each UID, or data item input by the user, define an attribute according to the following:
   - Verify that the data item is functionally dependent but not transitively dependent (see [4]) on the attribute OID in each class, i.e., that OID → data items. Verify that the data item is not transitively dependent on the OID. If these conditions are satisfied, then the data item must become an attribute of the class.
   - Verify that the data item is functionally dependent, but not transitively dependent, on the attribute OID of two or more different classes, and that the data item is not transitively dependent on the OIDs. If these conditions are satisfied, then the data item must become an attribute of a relationship between these classes.
3. For each attribute that appears in a set different from its class, tentatively define a relationship between its class and the class of the set elements. If the attribute class is not related to the class representing the set, then verify that the attribute class is related to the class of another attribute present in the set. Verify that this resulting

relationship is semantically correct (i.e. if it makes sense in the domain being modeled).

4. For each interaction flow (represented by an arrow), if there are different classes in the source interaction and the target interaction, define a relationship between these classes. Verify if this relationship is semantically correct (i.e. if it makes sense in the domain being modeled).

Fig. 6 shows the part of the class diagram that results from these steps, for the example.

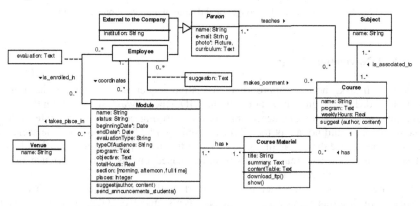

Fig. 6. Part of the conceptual schema of the application

3. Synthesis of the Navigational Model

3.1 Overview

The previous phases have produced, for each task, several scenarios, a use case and a UID. We describe next a method (first proposed in [7]) to derive the navigational topology of an application that supports each task, and then unify the navigational solutions of the different tasks into a general solution for the application. The designer may reuse solutions or apply design patterns, according to his own experience. The goal is to obtain the OOHDM Navigation Model of the application.

The main phases of the navigational design are shown in **Fig. 7** below; each of the sub-tasks is examined in more detailed in the next sub-sections.

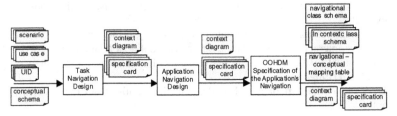

Fig. 7. Main phases of the navigational design, with their inputs and outputs.

3.2. Task Navigation Design

For each task, the most appropriate navigation sequence that supports the user is determined. The designer may start with the UID, since it was validated with the users and represents, therefore, an acceptable form of carrying out the task. In spite of this, the designer should not adopt the UID as its only guide, but also consult the scenario descriptions, since these contain the knowledge that users have about the task at hand. The designer can also reuse solutions either based on his own experience or by applying design patterns [11].

There are tasks where the user needs to work with sets of objects, which can be explored in different ways according to the user's objective. OOHDM has a design primitive for such sets, called *navigational context*. The navigational structure of an application is characterized as a group of contexts within which the objects will be accessed. This is specified in the context diagram, as exemplified in **Fig. 8**.

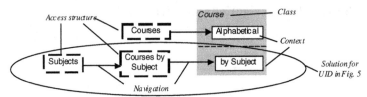

Fig. 8. Example of a context diagram. The indicated sub-diagram constitutes a possible navigational solution for the task described in the UID in Fig. 5.

The main steps in the task navigation design the are represented in the **Fig. 9**.

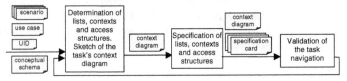

Fig. 9. Steps to design the navigation of a task

3.2.1 Determination of Lists, Contexts and Access Structures; Sketch of Individual Task Context Diagram

Each set represented in the UID is analyzed to determine the type of navigation primitive it will become: an access structure, a context or attribute of type list (an ordered set of elements). The solution will be represented in a context diagram.

The following rules can be applied during the analysis of each UID:

Rule 1: An interaction that presents an object is mapped into a context of its class

Objects are navigated within a context; therefore any interaction that presents an object should be mapped into a context. In **Fig. 5** the fourth interaction of the UID presents an object of the class Course, so it should be mapped into a context of this class.

The next step is to determine the elements of each context. The elements of a context generally have some common characteristic: they belong to the same class. or they have attributes with the same value, or they are linked to a given object through the same relationship. Otherwise, the set is arbitrary, with enumerated elements.

The "Course by Subject" context in **Fig. 10** represents the mapping of the interactions of the UID in **Fig. 5** that were analyzed according to rule 1.

There are several guidelines that may be used for identifying context elements, not shown here for reasons of space.

Rule 2: A set that is part of the information within an object is mapped into an access structure, or into a context, or into an attribute of type list

In the UID in **Fig. 5** the information about a course includes the sets: Subject (subjects associated with the course), Course Material and Teacher (people who teach the course). To decide the mapping of each of these sets we analyze the importance of its information for the execution of the task, based on the use case and scenarios.

- Very important information should appear integrally an object attribute of type list that contains the set elements;
- Important information may appear as an access structure within the object. In this case the designer should decide the attributes that will be shown in the structure; each access structure element will point to a set element;
- Complementary information may appear as an anchor pointing to an access structure or to an element within a context:

Fig. 10 represents the mapping of the UID in **Fig. 5** after the course sets have been analyzed. For each context, we include a card with its object vision and its access permissions. It may happen that the same object can be accessed in different contexts and it may be necessary that in each context the object presents the information in different ways, as well as allowing different navigations. Such variations will be solved through *InContext classes*.

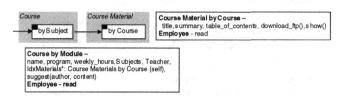

Fig. 10. Part of the context diagram of the use case "Browsing courses given a subject" (rule 2)

Rule 3 A data entry interaction followed by an interaction that presents a set of objects is mapped into an access structure pointing the objects in the set

The choice of the access structure depends on whether the possible values of the data entered by the user can be computed by the application in advance, and whether this data belongs to the objects of the target set or to other objects linked to them.

The UID in **Fig. 11** shows the task "search modules given a keyword". The user can enter a keyword, also selecting the attributes where it will be searched; the application returns the modules that match the search.

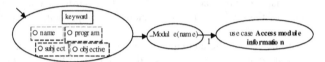

Fig. 11. User interaction diagram for the use case "Searching modules by keyword"

In this case, the system is not able to compute the values of the user's input beforehand. In such cases, the data entry interaction and the resulting set are mapped into an access structure whose elements will be computed at runtime from the user's input (*dynamic structure*). **Fig. 12** represents the mapping of **Fig. 11**'s UID:

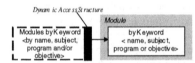

Fig. 12. Context diagram, use case "Searching modules by keyword"

In the UID in **Fig. 5**, the user must enter a subject, and the system returns the courses associated with that subject. Since in this case the application knows beforehand all possible valid inputs by the user, it could generate all the subjects stored in the knowledge base for the user to choose one, without having to actually enter it. Therefore, both interactions are mapped into an access structure. The type of the access structure, simple or hierarchical, will depend on the data entry and on the objects of the destination set. A hierarchical access structure is a sequence of simple structures where the selection of each structure is the input parameter for the succeeding one [13].

In the diagram in **Fig. 13**, the user enters a subject and the target set is formed with courses linked to this subject, forming a hierarchical access structure. In the first level, the subjects are presented ordered alphabetically, and in the second level the courses for each subject selected in the first level are presented. **Fig. 13** represents the complete mapping of the diagram represented in **Fig. 5**:

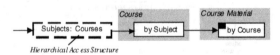

Fig. 13. Context diagram of the "Browsing courses given a subject", corresponding to the UID in Fig. 5. Levels in a hierarchical access structure are separated by the character ":"

3.2.2 Specification of Lists, Contexts and Access Structures

The construction of the context diagram of the tasks is completed with the specification of the lists, contexts and access structures in their respective cards. These cards record data that are not represented in the context diagram: the characteristics of the elements of the set, their ordering, the internal navigation topology, etc. A specification card is defined for each list that is an object attribute; for each context represented in the context diagram; and for each type of access structure used in the context diagram. This is true regardless of the structure being represented graphically or being described as an attribute of a class in some context vision.

3.2.3 Validation of Each Task Navigation

The task navigation is validated with the users. Context diagrams are used as a communication tool between designer and users. Based on interview records and annotations, and on his own experience, the designer will opt for a solution, trying to reconcile the changes proposed by the various users. If the validation results in significant changes, the context diagram is sketched again and the process is repeated. Otherwise, the designer can continue on to the next step of the process.

If the users present conflicting opinions about the solutions presented, this may actually indicate that different applications may be needed to accommodate this group of users.

3.3. Application Navigation Design

The application's context diagram will be synthesized through the successive union of the context diagrams of the individual tasks. For each union step, a new context diagram is built, defining its contexts, its access structures and the object visions for each context.

We start by unifying the solutions of each user group, starting with the group that performs the main tasks. Tasks in which the same navigational object is accessed are unified in sequence, as soon as each solution is defined.

The same navigational class may appear in contexts present in different partial diagrams, so we analyze if a single one may replace some of these contexts. We analyze the tasks that each context supports, its object visions and user permissions. If the resulting context supports all such tasks, and if the object visions and the user permissions do not conflict, we replace them by a single context

We also analyze the possibility of unifying the access paths leading to the original contexts and if it is possible to generalize the original navigations of these contexts within the resulting diagram. Since they are now part of the same diagram, it is necessary to explore the possible navigations between them.

As an example, we will unify some Employee tasks. We will unify first the solutions for the use cases "Searching modules by keyword" (**Fig. 12**) and "Browsing modules given a subject" (**Fig. 14**), since the class Module is accessed in both.

Fig. 14. Context diagrams for "Browsing modules given a subject" task

The contexts that are subsets of others may be both replaced by a single context. The context "Module by Subject" is included in the context "Module by Keyword". From the context "Module by Keyword" it is possible to access the modules of a certain subject, therefore "Module by Subject" could be replaced by "Module by Keyword".

After analyzing the tasks, we opt to maintain both contexts. Search based on a keyword is useful for users that know the module's characteristics, while navigation guided by access structures leading to "Module by Subject" is useful for beginners.

We must then analyze how will the joint navigation of these contexts be in the resulting diagram. According to the conceptual model, every module can have courses (context "Course by Module") associated to it. Evidently, this navigation can be generalized for all contexts of class Module, including "Module by Keyword" as shown in **Fig. 15**. Notice that the union with use case "Browsing courses given a subject" (**Fig. 13**) was also represented, so we unified all employee's tasks.

A similar analysis is made for the other tasks; the context diagram of **Fig. 15** supports all tasks that were previously discussed: "Browsing courses given a subject", "Browsing modules given a subject", "Searching modules by keyword", and "Browsing coordinator's module".

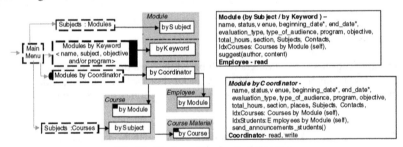

Fig. 15. Final resulting context diagram

3.4. Navigational Specification of the Application

The OOHDM navigational design specification is a set of models, enumerated here in the same order that the project documentation should be presented: one conceptual-navigational mapping table, one navigational classes schema, several In Context classes schemas, one context diagram, and several specification cards. The context diagram and the cards were obtained in the union process.

3.4.1 Navigational Class Schema

Navigational classes describe the information content through which the users will navigate. These classes constitute visions of conceptual classes: a navigational class can present some information of a conceptual class or a collage of contents of several conceptual classes. Navigational classes are called nodes; navigational relationships are called links; and node attributes that activate navigations are called anchors.

The **Fig. 16** shows the nodes and links in the navigational class schema for the example.

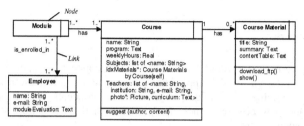

Fig. 16. Part of the navigational class schema of the application

3.4.2 InContext Class Schema

In this step we will build the InContext class schema for each node. We derive the In Context classes from the navigations and the object vision cards of the context diagrams, which record the peculiarities of the object in each context. The object peculiarities must be defined as "decorators" that enrich the object when it is visited within the different contexts, called *InContext classes* in OOHDM. The complete navigation object that the user navigates is the combination of the "pure" navigational object with its InContext Classes, according to each particular context within which the navigation takes place.

In the example, the only class that has several visions is Module: one for the students (contexts "by Subject" and "by Keyword") and another for the coordinators (context "by Coordinator"). Therefore only the Module's In Context classes schema is defined (**Fig. 17**).

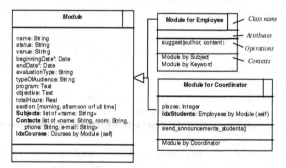

Fig. 17. Class Module In Context class diagram

3.5. Final Specification of the Application's Navigation

In this step, the specification cards are revised with the goal of eliminating any difference with the navigational model. For example, the context "Module by

Subject" was specified initially with the node "Subject", but at the end of the design; "Subject" wasn't defined as a navigational class. In fact, it was turned into an attribute of node Module, specified in the conceptual-navigational mapping table. The selection rule in the card therefore must be expressed only in navigational classes.

4. Conclusions

We have presented a method that allows requirements gathering, conceptual design and navigation design of Web applications, based on scenarios, use case and UIDs. This method combines, in a unique way, traditional concepts, such as scenarios and use cases, and new concepts, such as the UIDs, in the navigation of hypermedia application development.

Several approaches also use scenarios and use cases for gathering requirements, but in a different way. In [14], scenarios are used to validate the requirements and are automatically generated from use cases elicited from the users. The scenarios used in [5], different from our scenarios, describe interface and navigation aspects, since they are used in the redesign of an existing web site.

In [8] the requirements gathering is based on use cases derived from user stories. A user story can be seen as a simpler scenario, where an agent performs a physical action on an object. However, in both users stories and use cases, the interactions between the users and the system are specified only as a textual description.

In the Unified Process [10], the requirements are also defined by use cases, but scenarios are not used to specify them. If necessary, user interface prototypes can be used to understand and specify the requirements, but there are no diagrams that focus specifically on the interaction between the user and application.

The method WSDM [3], for designing read-only web sites, proposes a similar approach than the one presented here. In WSDM the navigation for each audience class is designed separately and the navigational model of the web site is the collection of disjoint navigation tracks of the several audience classes, linked by an initial element. Since different user groups could need to navigate through the same track, the navigational structure of a application should not be disjoint. In our approach different user groups can navigate through the same track, because OOHDM models permit specifying applications with complex navigations and behavior and not simply read-only.

Although OOHDM currently allows a simple form of navigation customization according to a user role, a more complete model is needed. We are currently developing an extension to OOHDM that will allow the specification of customized user views, based on user roles and identity.

The approach presented here has been used in several real-life applications being developed, with encouraging results. For the example mentioned in this paper, 23 users of six different classes were interviewed, eliciting 169 tasks in 364 scenarios, later filtered down by the project managers into 63 tasks and 128 scenarios. The final application has 39 UIDs, resulting in 61 conceptual classes, 29 navigation classes, 47 InContext classes, 45 contexts, and 41 access structures.

The formal validation of this approach type is extremely difficult, but it was observed that, besides the easy adoption by the users in the validation process, it was possible to execute applications project in smaller time, and with fewer revisions than in previous projects of similar dimensions.

5. References

1. Carroll, J.M. Scenario-Based Design: Envisioning Work and Technology in System Development, John Wiley & Sons, 1995.
2. Booch, G., Rumbaugh, J., Jacobson, I.: The Unified Modeling Language User Guide. Addison-Wesley, Reading, 1999.
3. De Troyer, O. Audience-driven Web Design, in Conceptual Modeling in the Next Millennium, Rossi, M. (ed.), CRC Press, USA, 2000
4. Elmasri, R., Navathe, S.B., Fundamentals of Database Systems, Benjamin Cummings, 1994.
5. Erskine, L.E., Carter-Tod, D.R.N., and Burton, J.K. Dialogical techniques for the design of web sites. Int. Journal Human-Computer Studies, 47 (1997), 169-195.
6. Garzotto, F., Paolini, P., and Schwabe, D. HDM - A Model-Based Approach to Hypertext Application Design. ACM Transactions on Information Systems 11, 1, 1993, 1-26.
7. Güell, N. User Centered Navigation Design of Hypermedia Applications (in Portuguese). Tech. Report MCC10/98, Departamento de Informática, PUC-Rio (1998). 18p. Available in ftp://ftp.inf.puc-rio.br/pub/docs/techreports/98_10_schwabe.pdf.gz
8. Imaz, M., and Benyon, D. How Stories Capture Interactions, in Proceedings of Human-Computer Interaction - INTERACT'99, IOS Press, 321-328.
9. Isakowitz, T., Stohr, E., and Balasubramanian, P. RMM: A methodology for structuring hypermedia design. Communications of. ACM 38, 8 (August 1995), 34-44.
10. Jacobson, I., Booch, G., and Rumbaugh, J. The Unified Software Development Process, Addison-Wesley, 1999.
11. Rossi, G.; Schwabe, D.; Lyardet, F.; "Improving Web Information Systems with Navigational Patterns", The International Journal of Computer and Telecommunications Networking, Elsevier, May, 1999 pp. 589-600.(ISBN 0-444-50264-5)
12. G. Rossi, D. Schwabe, F. Lyardet: "Web application models are more than conceptual models". Proceedings of the First International Workshop on Conceptual Modeling and the WWW, Paris, France, November 1999, Lecture Notes in Computer Science, Vol. 1727, Springer, 1999, 239-253.
13. Schwabe, D., and Rossi, G. An object-oriented approach to Web-based application design. Theory and Practice of Object Systems, v. 4#4, (October 1998), 207-225.
14. Sutcliffe, A.G., Maiden, N.A.M., Minocha, S., and Manuel, D. Supporting Scenario-Based Requirements Engineering. IEEE Transactions on Software Engineering 24, 12 (December 1998), 1072-1088.
15. Vilain, P., Schwabe, D., de Souza, C. S.: A Diagrammatic Tool for Representing User Interaction in UML. To appear in UML 2000 -Third International Conference on the Unified Modeling Language, (York, UK, October 2000).

A General Methodological Framework for the Development of Web-Based Information Systems*

Silvana Castano[1], Luigi Palopoli[2], and Riccardo Torlone[3]

[1] DSI, Università di Milano, Italy. castano@dsi.unimi.it
[2] DEIS, Università della Calabria, Italy. palopoli@si.deis.unical.it
[3] DIA, Università di Roma Tre, Italy. torlone@dia.uniroma3.it

Abstract. In this paper, we present a general methodological framework, called WISDOM (Web Based Information System Development with a cOmprehensive Methodology), for the development of Web-based information systems (WIS). WISDOM is generally applicable and supports the design and the development of a wide spectrum of WIS applications. This is achieved by defining the WISDOM framework as a family of inter-related activities. The most suitable combination of activities for the development of a target WIS application is obtained by dropping and/or specializing one or more of the component activities of WISDOM.

1 Introduction

In the last few years, an important growth of Web-based approaches to the development of information systems has been witnessed within very diverse application scenarios. Accordingly, there has been a significant proliferation of technologies supporting developers of Web-based information system (WIS). Moreover, the development of new tools and techniques for WIS management represents nowadays one of the most relevant issues for theoriticians and practitioners in the area. As an example, several tools equipped with high-level interfaces have been defined and implemented for retrieving and processing information extracted from the Web and/or to build Web sites over data repositories organized in the most proper way for this purpose [8,12].

Unfortunately, methodological issues pertaining Web-based information system design and development have not received much attention, even though the development of WIS applications is usually a complex task, often involving the solution of quite diverse problems. Actually, WIS applications may vary significantly from one another, both in terms of application objectives and in terms of architecture and structure and also of supplied functionality. This makes the development of a methodological framework spanning a WIS life-cycle, complex to define. In particular, with WIS applications, aspects related to distribution,

---

* This research has been partially supported by MURST, within the *InterData* project, and by CNR.

S.W. Liddle, H.C. Mayr, B. Thalheim (Eds.): ER 2000 Workshop, LNCS 1921, pp. 128–139, 2000.

heterogeneity, and availability of information sources as well as to workflow processes are important and have to be considered. In the development of WIS applications, both *information export activities* (devoted to the organization and publication of information over the Internet/Intranet), and *information import activities* (devoted to the collection, extraction, and integration of information available in existing Web information sources for transaction and analytical processing purposes) are to be considered into account and properly inter-related.

The goal of this paper is the definition of a methodology general enough to effectively support the development of complex and diverse WIS applications in which the contents of information sources are analyzed, described, extracted, integrated, processed and possibly reorganized in order to create new information sources. To this aim we propose a *structured methodological framework*, called WISDOM (Web Based Information System Development with a cOmprehensive Methodology) that can be properly customized to a specific WIS application development. WISDOM is made of a large number of activities, independent of each other but strongly coordinated. The whole scheme is structured in that complex activities are iteratively refined in a number of sub-activities. Once the application requirements are known, a specific methodology for the development of the target application can be identified in WISDOM by selecting the most suitable sequence of activities. In this sense, WISDOM provides a framework of reference for the development of *generic* WIS applications: a "view" or "instance" of WISDOM can be obtained by dropping and/or specializing one or more of its component activities, and constitutes a specific methodology for a given application.

The work reported in this paper has been developed within a national Italian project, called InterData, whose aim is to study and develop methodologies and techniques for managing data in Web environments. Some of the issues presented in this paper have been therefore largely influenced by the results of this project. In more detail, Web site design borrows ideas from [1,11], workflow design from [2,9], data integration from [6,7,13,14] and, finally, warehousing from [5].

Providing an integrated framework where both information export and import activities are put together and coordinated is an important step for WIS application development. The effort we made in this respect constitutes the novel contribution of the work. To the best of our knowledge, this is a first paper addressing the development of generic WIS applications with such a comprehensive view of both information import and export activities.

The paper is organized as follows. In Section 2, we settle down the general WISDOM framework. In Section 3, we analyze the activity of information export, whereas in Section 4, we focus on information import. Then, in Section 5, we describe the customization of WISDOM to target WIS applications.

2   The WISDOM Framework

The general structure of WISDOM is illustrated in Figure 1. In the figure, (*i*) a box represents an activity, possibly further decomposed into sub-activities; (*ii*) a link between two activities represents both an order relationship between their execution and a flow of information; *(iii)* free texts represent input/output data expected/produced by an activity, as indicated by an arrow.

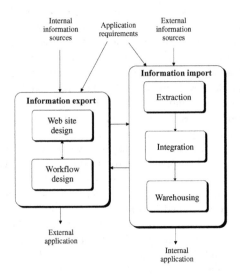

Fig. 1. The WISDOM methodological framework

According to what we have said in the introduction, WISDOM considers two basic activities in the context of Web applications: information *import* and information *export*.

Information export. This activity is devoted to the organization and publication of information over the Internet/Intranet and Web, and possibly to the design of workflow processes over the Web. Published information can originate from (possibly pre-existent) internal data sources or can originate from the integration of proprietary information with data extracted from external data sources. The result of this activity is an *external application* in the sense that it can be accessed through the Web. Information export is decomposed into the following macro-activities (see Fig. 1): *Web site design*, where data to be published are properly organized and managed, possibly using a database management system, and presentation and navigation features of the Web-interface are defined; *Workflow design*, where processes and services provided by the WIS under development are identified, and basic tasks of these processes are defined and coordinated using workflows. If both activities are required, they often need to be executed in parallel since they influence each other.

Information import. Information import consists in the collection, extraction and integration of relevant information available in existing Web sites or data sources for transaction and analytical processing purposes. As shown in the picture, the result of this can be an *internal application*, that is, an application available in a local network/intranet within the organization. Information import is decomposed into the following main activities (see Fig. 1): *Extraction* of the information of interest from external information sources through wrappers; *Integration* of extracted data, to provide a unified representation of heterogeneous information; *Warehousing* which consists of a further processing of integrated data, mainly based on multidimensional analysis and usually oriented to decision support. Differently from the case of Web site and workflow design, these activities proceed in sequence since each of them needs the outcomes of the preceding activity. As often happens however, a feedback among phases is required to improve and correct results of the previous phases.

Note that the development of a target WIS application often requires both import and export activities. For instance, an application could require the integration of internal data with data extracted from the Web. Integrated data can can be then processed within an internal application or an external one.

In the following sections, we will describe in a top-down way, the various activities reported in Figure 1, resulting in the sub-activities depicted in Figure 2.

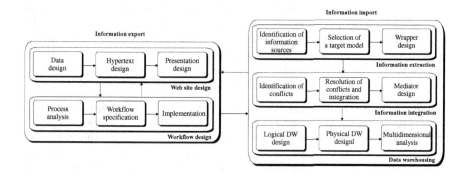

Fig. 2. The WISDOM methodological framework - Refined view

3   Information Export in WISDOM

The Web is rapidly becoming a standard interface to information systems and, more in general, a uniform platform for sharing data, and traditional methodologies for the design of information systems must be reconsidered in this context. The first goal of WISDOM is to provide a support to the development of modern information systems that need to publish and exchange data using the Web. Two major activities can be identified in this context: Web site design and workflow design. In the former, the Web site and the underlying database used as back-end are defined. In the latter, services and applications are developed to support business process execution over the Web.

3.1 Web Site Design

In the Web scenario, information systems usually maintain large amounts of well structured data that are stored in database management systems and are accessed through a Web site. In WISDOM, the main objectives of Web site design are (a) associating a Web site with a high-level description of its content, that can be used for querying, evolution, and maintenance; and (b) separating the information content, stored in the database from navigation and presentation features, which can be independently defined. Thus, the methodology distinguishes three main sub-activities: the data design, the hypertext design, and the presentation design (see Figure 2).

Data design. This phase is required when the Web based information system is built from scratch. For data design, we follow a traditional approach to database design based on a separation between a conceptual and a logical phase [4]. The output of the first phase is a conceptual scheme (E-R or O-O), describing the organization of the underlying domain in an implementation-independent manner. From the conceptual scheme, the logical (and possibly physical) scheme of the database can be derived using standard techniques. It may happen that the database containing the information to publish is already existent. In this case, data design can be omitted, provided that a conceptual scheme is available. If this is not the case, a reverse-engineering step may be necessary in order to obtain a conceptual scheme starting from the existing database.

Hypertext design. In order to better isolate the structural properties of the resulting hypertext, it is useful to consider two different description levels: the hypertext conceptual level and the hypertext logical level. At the conceptual level, the hypertext is simply described in terms of nodes and paths to navigate between them. This can be done by using a specific data model inspired by known hypertext data model. At the logical level the hypertext organization is detailed in terms of Web pages and links. This can be done using a logical model for Web hypertexts [1]. The hypertext conceptual design aims at describing how application domain concepts can be organized in hypertextual form. This activity is independent of the physical implementation, and concentrates on two essential aspects: (i) deciding which concepts (or combination thereof) of the input scheme will correspond to hypertextual nodes; (ii) choosing the paths to navigate between concepts. The hypertext logical design describes the actual organization in terms of pages and links of the Web site. This phase is specific of the Web framework, and aims at detailing the structure of the hypertext as shown by the browser. At this level we concentrate on abstracting the relevant pieces of information in the page (e.g., text or images) and their organization (e.g., at page, list or nested list). To do this, the hypertext conceptual scheme is translated into a logical scheme, based on a suitable logical data model for hypertext applications providing a notion of *page type* and where each page is seen as an object with an URL plus a set of attributes. Page types are connected using links, used to describe navigation in the site.

Presentation Design. In the Presentation Design, the final graphical layout of Web pages is defined. The input of this phase is the logical hypertext description of the Web site and the output are a collection of Web pages in the form of HTML documents or XML files plus XSL style sheets. We associate a page style with a page type of the hypertext logical schema. Such a page style specifies all format directives for each piece of information in the page, plus the graphical features to be associated with the page itself, like, for example, page background and banners. This can be done using an abstract description or a standard language like XSL, whereby Web pages can be automatically derived [11].

3.2   Workflow Design

Workflow design in WISDOM pertains to all the activities to be carried out in order to build applications supporting business process execution over the Web. As an example of business process execution over the Web, consider an e-commerce process for sales. Its description consists of information about the involved activities and information objects (e.g., selection of items to purchase, filling of a form, checking fund availability for a credit card, etc.), the involved agents (e.g., seller, buyer), and the business goals (e.g., high-level perceived security). This process can be naturally modeled in the form of a workflow. The activities composing the workflow design macro-activity in WISDOM are described in the following (see Figure 2).

Process analysis. Process analysis starts from a description of the business process to be implemented covering the following perspectives [2]: (i) functional, concerning the process activities and the involved information objects; (ii) organizational, concerning the agents and roles involved in process execution; (iii) business, concerning the goals capturing business rules and objectives of the process. A workflow is suggested for each group of activities that are loosely coupled with the outside and have a high number of connection points with activities within group, and which are performed within different organizational units. For each candidate workflow, roles, pre-conditions (i.e., the event(s) starting the workflow), post-conditions (i.e., how the workflow ends), and associated business goals, are specified.

Workflow specification. The goal of this activity is the definition of a schema for each candidate workflow according to a conceptual workflow model [2,9]. According to such a model, a workflow schema is composed of sub-processes and tasks. Tasks are organized into a directed graph, which defines their execution order. Arcs in the graph can be labeled with transition predicates defined over process data, with the meaning that the tasks connected through outgoing arcs are executed only if the corresponding transition predicate evaluates to true. A top-down development is assumed in the workflow design phase. A candidate workflow resulting from the process analysis phase is decomposed into sub-processes and then into tasks, which are properly inter-related into a flow structure, to reflect the correct sequencing of activities in the sub-process. In a WIS application, sub-processes can describe self-contained activity fragments to be executed, for example, at a given site. Interactions with external information systems/applications are also modeled during this phase, by specifying at a conceptual level the expected interaction modalities between the elements of the workflow schema and external information systems/applications.

Workflow implementation. The final activity of workflow design in WISDOM consists in implementing the designed workflow, either by developing ad hoc software solutions or by using a commercial workflow management system.

4   Information Import in WISDOM

In order to run desired analysis and manipulation procedures over information coming from one or more existing sources, information must be first located and extracted from world-spread information source sites. Information coming from different sources can be heterogeneous and possible conflicting situations have to be identified and resolved to come up with an integrated, unified information representation. Moreover, integrated information may not have the right format to undergo needed decision-oriented analysis, and a warehousing activity can be required. Consequently, the goal of WISDOM for the information import activity is to provide a methodological support to guide the user in the activities of extraction, integration and warehousing.

4.1   Extraction

The extraction macro-activity in WISDOM pertains all the activities to be carried out in order to import the information at the site where analysis procedures take place. Obtaining information means either to materialize data at the application site (materialized approach) or maintaining suitable query templates to be executed to extract the information on demand (virtual approach) [10]. The extraction activity requires the identification of external information sources where relevant information to be extracted can be found. A target data model is selected to represent extracted information uniformly in view of subsequent integration activity, and wrapper tools are developed to perform data restructuring according to the selected target model. Hence, the Extraction macro-activity is refined as described in the following (see also Figure 2).

Identification of information sources. With this activity, application requirements must be analyzed to single out required information content as target elements. A *target element* refers to a single concept or to a restricted number of related concepts of interest for the application. Moreover, it provides a textual description of the information to be retrieved for it (e.g., *Employee* target element, with information regarding employee name, address, salary, qualification). The designer must identify all concepts relevant to the application, and define appropriate target elements for representing them. Potential information sources where data sets for the target elements required by the application can be located are established. Since we deal with WIS applications, information sources where to find data sets could be not known a priori, due to the lack of sufficient information on contents and location of the sources worldwide. In such a case, the entire Web is considered as the default information source, and the selection of most specific information sources is demanded to the subsequent phases.

Selection of a target data model. When retrieved from multiple, existing information sources, data sets can be formatted in different ways and may be not ready to be used for the purposes of developing the target WIS application. Therefore, the designer has to define a target data model, according to which extracted data sets will be restructured for the integration activity. In particular, the target data model can be a database model, or a semi-structured data model. After defining the target data model, the designer gives indication also of the storing data structures and presentation formats for schema elements.

Wrapper design. Data sets can be extracted from a WIS using various techniques. This include specialized query languages, search robots and custom procedures [12]. A WIS application development will typically require to choose more than one of such tools in order to locate and retrieve all the needed data sets for each element. It is important to stress that performing retrieval over the Web may also serve the purpose of data location. Moreover, note that a correct choice of retrieval tools is a key issue if efficiency is to be achieved, specially if "on-the-fly" retrieval of data is involved. For each target element, one or more query templates are defined to import the corresponding data sets. A query template specifies the location where to retrieve data (in the default case, the Web) and the extraction strategy. For materialized elements, query templates are executed and their results (i.e., data sets) are gathered according to the data format they have in the original source. As for virtual elements to be retrieved "on-the-fly", query templates are maintained in form of stored procedures at the local source. A wrapper is designed for each different data set to generate its corresponding representation according to the target data model, and vice versa. As the result of the wrapping, we obtain the so called *restructured data sets* are obtained to be used in the subsequent integration step.

4.2   Integration

This activity serves the purpose of resolving conflicts between restructured data sets obtained in the previous phase for a given target element. The goal is to obtain an integrated, unified representation of various target elements, and make them suitable for subsequent elaboration. Sub-activities composing the integration macro-activity are described in the following (see Figure 2).

Identification of conflicts. This activity is concerned with pointing out conflicts among restructured data sets. Conflicts are due to the fact that different external information sources may use different terminologies and design structures to describe the same concept. Using the target data model and wrapper tools facilitates conflict identification and resolution, in that all data representations are reduced to the same model. Following [3,10], we distinguish the following main categories of conflicts: (*i*) *lexical conflicts*: due to a different terminology adopted to denote a certain concept; (*ii*) *structural conflicts*: due to the use of different data structures used for representing a certain concept; (*iii*) *semantic conflicts*: due to the differences between utilized domain values (e.g., format, currency, unit). In addition, restructured data sets should be analyzed to discover also possible application-dependent conflicts. Semi-automatic techniques can be employed to assist the designer in the conflict identification activity [7,14].

Resolution of conflicts and integration. Given the list of conflicts identified in the previous step, this activity deals with conflict resolution, to obtain an integrated definition of the target application schema. Conflict resolution is a process heavily custom-interactive, since (specially when semi-structured data are involved), it is largely application-specific. However, semi-automatic techniques have been developed to support the designer in deriving the integrated schema out of restructured data sets plus conflicts [6,13]. As the result of the conflict resolution activity, the final target application schema is obtained.

Mediator design. The goal of this activity is to define mappings and conversion functions to materialize each target element by mediating the different structures of underlying restructured data sets. Given a target element, mappings defined for it specify the correspondences between its structure and the structure of restructured data sets from which it has been derived. Conversion functions are defined for attributes of the target element to implement transformations to convert mismatching domain values of its corresponding data sets, if necessary. In case of materialized target elements, defined mappings and conversion functions are executed on data sets to populate the target application schema. In case of virtual target elements, mapping and conversion functions are maintained together with the target application schema, and will be used for populating target elements "on-the-fly", in combination with stored query templates. Mappings and conversion functions together with the target application schema constitute the so called mediator module. Mediator functionality coupled with wrapping translation functionality previously described enforce the integration of heterogeneous data sets.

4.3 Warehousing

This macro-activity is concerned with the construction of an integrated collection of operational data, called *data warehouse*, followed by the processing of its content, usually oriented to decision making. The warehousing macro-activity of WISDOM consists of the following sub-activities (see Figure 2).

Logical data warehouse design. In this phase, the structure of the data warehouse is defined starting from an integrated schema. The integrated schema may have a format that is often not suitable for the analysis purposes. Therefore, a first step of the data warehousing process consists in the transformation of the input scheme in order to provide a better support for analysis operations. The output consists in a schema of the data warehouse according to a logical data model, which describes the multidimensional aspects of data analysis and is independent of the implementation in a specific data storage system. We make use of a logical data model for multidimensional databases [5]. The logical data warehousing design follows a structured approach consisting of a number of activities. The first activity consists in a careful analysis of the given E-R scheme whose aim is the selection of the facts, the measures, and the dimensions of interest for our business processing. The second activity consists in a reorganization of the original E-R scheme in order to describe facts and dimensions in a better, more explicit way. The goal of this step is the production of a new E-R scheme that can be easily translated into the logical multidimensional model. A dimensional graph is used to represent, in a succinct way, facts and dimensions of the restructured E-R scheme. The final activity consists in translating the dimensional graph into the logical multidimensional model.

Physical data warehouse design. In this phase the input schema is translated into the data model adopted by the storage system chosen. Since the multidimensional database is defined according to a logical data model, it can be in fact implemented in several ways (e.g., using either ROLAP or MOLAP systems). In the first case a multidimensional database can be implemented in the form of a "star scheme" (or variant thereof, e.g., snowflake scheme). In the second case, an f-table is represented by a n-dimensional matrix, storing each measure corresponding to a certain symbolic entry in the cell having the corresponding physical coordinates. A dimension can be then represented by means of a special data structure, with a hierarchical organization according to the hierarchy defined on it. We can then use this structure as an index to access the multidimensional array.

Multidimensional analysis. In this last step the data warehouse is finally used to perform the analytical processes for which it was originally designed. This can done by using specific OLAP tools having querying and reporting capabilities. Aggregate views of the warehouse can be materialized to support analysis needs. In this phase, the propagation to the warehouse of updates on source data is periodically required. The outcomes of the analysis can be published through the Web site: this clearly requires some of the activities described in Section 3.

5   Application of WISDOM

In order to pinpoint the effective applicability of our methodological framework, we will describe possible instantiations/customizations with respect to significant practical examples of applications.

Development of a data analysis application with a Web site interface. This is an example of *mixed* application, where WIS design involves aspects related to both information import to design an integrated warehouse supporting expected types of analysis by selecting and integrating pre-existing data sources and information

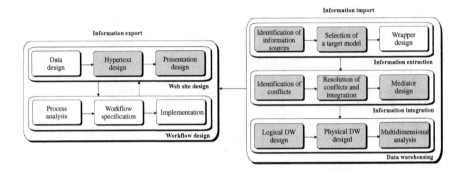

Fig. 3. Example of customized view of WISDOM

export, to design a Web site on top of the warehouse, to provide a common interface for querying the warehouse and publish analysis results. In this case, WISDOM can be customized by selecting the activities related to extraction, integration and warehousing. Moreover, for each macro-activity, only its sub-activities necessary in the specific application scenario are kept. For example, a possible customization of the methodological framework is shown in Fig. 3.

Design of an integrated schema of heterogeneous databases. This is an example of information import, where the goal is to define an integrated mediator schema of pre-existing heterogeneous databases to support uniform queries at the global level. In this case, WISDOM is customized by selecting the activities related to extraction and integration. Also in this case, some sub-activities can be skipped if not necessary.

Design of a Web-based document management workflow. This is an example of information export, where WIS design involves aspects related to coordination of activities across distributed organization units together with correct document exchanges among the involved units. This kind of application is typical, for example, of the Public Administration domain, where document management workflows are required to automate and make more efficient administration processes involving many organization units, often distributed over the territory. In this case, WISDOM can be customized by keeping only the activities related to workflow design. The emphasis is on designing activity and document flows correctly, by identifying all involved organization units, their responsibilities and interconnection modalities, and their expected interactions.

6   Concluding Remarks

In this paper, we have presented the WISDOM methodological framework for the development of WIS applications, conceived as a family of inter-related activities to support both information export and import activities. Future research work will be devoted to set out a complete instantiation scenario for a real case study of WIS application development. In this paper, we discussed the applicability of WISDOM by referring to the general typologies of WIS applications of Section 5. Specific case studies related to web site design, workflow design, integration, and warehousing activities have been separately studied in the framework of the Interdata project (see, for example, [6,13,5]), and are under consideration to develop a comprehensive case study. Another goal of future research work will be the enrichment of the framework with a set of pre-defined "methodological patterns", suggesting to the WIS designer a reference methodology for a specific application to be developed, that can be customized if necessary, based on the specific requirements of the application at hand.

References

1. P. Atzeni, G. Mecca, and P. Merialdo. Design and maintenance of data-intensive Web sites. In *Sixth Int. Conf. on Extending Database Technology (EDBT'98)*, 1998.
2. L. Baresi, F. Casati, S. Castano, M.G. Fugini, I. Mirbel, B. Pernici. WIDE Work-flow Development Methodology. In *Int. Joint Conf. on Work Activities Coordination and Collaboration (WACC'99)*, 1999.
3. C. Batini, M. Lenzerini, S.B. Navathe. A Comprehensive Analysis of Methodologies for Database Schema Integration. *ACM Computing Surveys*, Vol.18, No.4, 1986.
4. C. Batini, S. Ceri, and S. Navathe. *Conceptual Database Design: an Entity-Relationship Approach*. Benjamin & Cummings, 1992.
5. L. Cabibbo, R. Torlone. A logical approach to multidimensional databases. In *Sixth Int. Conf. on Extending Database Technology (EDBT'98)*, 1998.
6. S. Castano, V. De Antonellis, S. De Capitani Di Vimercati. Global Viewing of Heterogeneous Data Sources. *IEEE Trans. on Knowledge and Data Engineering*, to appear.
7. S. Castano, V. De Antonellis. A Discovery-Based Approach to Database Ontology Design. *Distributed and Parallel Databases*, Vol.7, N.1, 1999.
8. A. Deutsch et al. XML-QL: A Query Language for XML. World Wide Web Consortium, Working paper, *(http://www.w3.org/TR/NOTE-xml-ql)*, 1998.
9. P. Grefen, B.Pernici, G. Sanchez, (eds.), *Database Support for Workflow Management, The WIDE Project*. Kluwer Academic Publishers, 1999.
10. R. Hull. Managing Semantic Heterogeneity in Databases: A Theoretical Perspective. Tutorial presented at *PODS'97*, 1997.
11. G. Mecca, P. Merialdo, P. Atzeni, V. Crescenzi. The Araneus Guide to Web Site Development. In *ACM SIGMOD Workshop on the Web and Databases (WebDB'99)*, 1999.
12. A. Mendelzon, G. Mihaila, T. Milo. Querying the World Wide Web. In *First Int. Conf. on Parallel and Distributed Information Systems (PDIS'96)*, 1996.
13. L. Palopoli, L. Pontieri, G. Terracina, D. Ursino. Intensional and Extensional Integration and Abstraction of Heterogeneous Databases. *Data and Knowledge Engineering*, to appear.
14. L. Palopoli, D. Saccà and D. Ursino. Semi-automatic, semantic discovery of properties from database schemes. In *IDEAS'98*, 1998.

Managing RDF Metadata for Community Webs[*]

Sofia Alexaki[1,2], Vassilis Christophides[1], Gregory Karvounarakis[1,2],
Dimitris Plexousakis[1,2], Karsten Tolle[3], Bernd Amann[4], Irini Fundulaki[4],
Michel Scholl[4], and Anne-Marie Vercoustre[4]

[1] ICS-FORTH, Vassilika Vouton, P.O.Box 1385, GR 711 10, Heraklion, Greece
{alexaki, christop, gregkar, dp}@ics.forth.gr
[2] Department of Computer Science, University of Crete, GR 71409, Heraklion, Greece
{alexaki, gregkar, dp}@csd.uoc.gr
[3] Johann Wolfgang Goethe-University, Robert-Mayer-Str. 11-15, P.O.Box 11 19 32,
D-60054 Frankfurt/Main, Germany
tolle@dbis.informatik.uni-frankfurt.de
[4] INRIA Rocquencourt, 78153 Le Chesnay Cedex, France
{amann, fundulak, scholl, vercoust}@cosmos.inria.fr

Abstract. The need for descriptive information, i.e., metadata, about
Web resources has been recognized in several application contexts (e.g.,
digital libraries, portals). The Resource Description Framework (RDF)
aims at facilitating the creation and exchange of metadata, as directed
labeled graphs serialized in XML. In particular, the definition of schema
vocabularies enables the interpretation of semistructured RDF descriptions
using taxonomies of node and edge labels. In this paper, we propose
(i) a formal model capturing RDF schema constructs; (ii) a declarative
query language featuring generalized path expressions for taxonomies of
labels (iii) a metadata management architecture for efficient storage and
querying of RDF descriptions and schemas.

1 Introduction

Metadata are widely used in order to fully exploit information resources (e.g.,
sites, documents, data, etc.) available on the WWW [13]. Indeed, metadata
permit the description of the content and/or structure of WWW resources in
various application contexts: digital libraries, infomediaries, enterprise portals,
etc. The Resource Description Framework (RDF) [21] aims at facilitating the
creation and exchange of metadata, as any other Web data. More precisely, RDF
descriptive (meta)data are represented as *directed labeled graphs* (where nodes
are called *resources* and edges are called *properties*) which are serialized using
an XML syntax. Furthermore, RDF schema [7] vocabularies are used to define
the labels of nodes (called *classes*) and edges (called *property types*) that can be
used to describe and query resources in specific user communities.These labels
can be organized into appropriate taxonomies, carrying inclusion semantics. In

---

[*] This work was partially supported by the European project C-Web (IST-1999-
13479).

S.W. Liddle, H.C. Mayr, B. Thalheim (Eds.): ER 2000 Workshop, LNCS 1921, pp. 140–151, 2000.

this paper, we are focusing on the design of a metadata management system for storing and querying both RDF descriptions and schemas as semistructured data [2].

Our work is motivated by the fact that existing semistructured models (e.g., OEM [23], YAT [12,11]) cannot capture the semantics of node and edge labels provided by RDF schemas (i.e., taxonomies of classes and property types), while semistructured or XML query languages (e.g., LOREL [4], UnQL [8], StruQL [17], XML-QL [15], XML-GL [10]) are not suited to exploit taxonomies of labels for query evaluation and optimization (i.e., pattern vs. semantic matching of labels). On the other hand, schema query languages as SchemaSQL [20], XSQL [19] or Noodle [22] do provide facilities for querying both schema and data. However, since they are based on common (relational/object-oriented) data models, they also fail to fully accomodate RDF/RDFS features - such as specialization of properties - and also impose strict typing on the data. In this context, we propose *RQL*, a declarative query language for RDF. *RQL* relies on a graph data model allowing us to (partially) interpret semistructured RDF descriptions by means of one or more RDF schemas. Thus, *RQL* adapts the functionality of semistructured query languages to the peculiarities of RDF but also extends this functionality in order to query RDF schemas.

The remainder of this paper makes the following contributions: Section 2, introduces a graph data model capturing RDF schema constructs [21,7]. The originality of our model lies on the distinction between classes and relationship types in the style of ODMG [9], as well as in the introduction of a graph instantiation mechanism, inspired by GRAM [6]. Section 3, presents the *RQL* language for querying semistructured RDF descriptions and schemas. *RQL* adopts the syntax and functional approach of OQL [9] while it features generalized path expressions in the style of POQL [3]. The novelty of *RQL* lies in its ability to query complex semistructured (meta)data and schema graphs using - in a transparent way - taxonomies of labels. Section 4 illustrates how we can benefit from schema information in order to validate and efficiently store RDF descriptions in a DBMS. Finally, section 5 presents conclusions and discusses further research.

2 Towards a Formal Model for RDF

In this section, we briefly recall the main modeling primitives proposed in the RDF Model & Syntax and Schema specifications [21,7] and introduce our graph model (for more details see [18]).

RDF schemas are used to declare *classes* and *property-types*, typically authored for a specific community or domain. The upper part of Figure 1 illustrates such a schema for a cultural application. The scope of the declarations is determined by the *namespace* of the schema, e.g., *ns*1 (http://www.culture.gr/schema-.rdf). Classes and property types are uniquely identified by prefixing their names with their schema namespace, as for example, *ns*1#Artist or *ns*1#creates. To simplify our presentation, we hereforth omit the namespace prefixes and denote by C the set of class names and by P the set of property types defined in a schema. Moreover, classes can be organized into a taxonomy through *simple* or *multiple* specialization. The root of this hierarchy, is a built-in class called `Resource`. For instance, `Painter` and `Painting` are subclasses of `Artist` and `Artifact` respectively, both specializing `Resource`. RDF classes do not impose any structure to their objects and class hierarchies simply carry inclusion semantics.

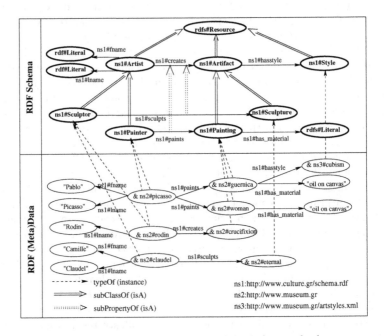

Fig. 1. An example of semistructured RDF data and schemas

RDF property types serve to represent *attributes* of resources as well as *relationships* (or *roles*) between resources. For example, *creates* defines a relationship between the resource classes `Artist` (its domain) and `Artifact` (its range) while *fname* is an attribute of `Artist` with type `Literal`.[1] As we can see in Figure 1, property types may also be refined: *paints* is a specialization of *creates*, with its domain and range restricted to the classes `Painter` and `Painting`, respectively. We denote by $H = (N, \prec)$, a hierarchy of classes and property types, where $N = C \cup P$. H is *well-formed* if $\prec$ is a smallest partial ordering such that :

- if $c \in C$ then $c \prec Resource$ (i.e., the root of the class hierarchy).
- if $p_1, p_2 \in P$ and $p_1 \prec p_2$ then $domain(p_1) \prec domain(p_2)$ and $range(p_1) \prec range(p_2)$.

---

[1] As RDF literals we can have any primitive datatype defined in XML as well as XML markup which is not further interpreted by an RDF processor.

Besides literal and property types, RDF also supports *container* types, i.e., **Bag**, **Sequence** or **Alternative**. Members of containers are identified by a unique integer index label i, while no restriction is made on their types (i.e., may have heterogeneous member types). RDF classes and container types correspond to schema graph nodes whereas property types correspond to edges.

Definition 1. *An RDF schema is a directed labeled graph* $RS = (V_S, E_S, \psi, \lambda, H)$ *where,* V_S *is the set of nodes and* E_S *is the set of edges,* $H = (N, \prec)$ *is a well-formed hierarchy of classes and property types (including Bag, Seq, Alt, Literal),* λ *is a labeling function* $\lambda : V_S \cup E_S \to N$, *and* ψ *is an incidence function* $\psi : E_S \to V_S \times V_S$, *capturing the* domain *and* range *of properties.*

In RDF, *Resources* are described through a collection of *Statements* committing to a schema (see lower part of Figure 1). As a resource we consider anything identifiable by an URI: it may be a Web page (e.g., http://www.museum.gr/picasso.htm), a fragment of a Web page (e.g., http://www.museum.gr/artstyles.xml-#cubism) or an entire Web site (e.g., http://www.museum.gr). In the sequel, we denote by O the set of resource identifiers composed by a *namespace* and a file name or anchor id (e.g., &ns2#picasso, &ns3#cubism). A non-disjoint population function π assigns to each class c in C a set of object identifiers $\pi(c)$, such that: $\cup\{\pi(c') \mid c' \prec c\} \subseteq \pi(c)$.

A specific resource together with a named property and its value form an RDF statement, represented by an ordered pair $< v_1, v_2 >$, where v_1 is its subject and v_2 is its object. The subject (e.g., &ns2#picasso) and object (e.g., "Pablo") should be of a type compatible (under class specialization) with the domain and range of the used predicate (e.g., fname). Figure 1 shows that RDF properties can be *multi-valued* (e.g., two *paints* properties for &ns2#picasso), *optional* (e.g., there is no *fname* property for &ns2#rodin) and they can be inherited (e.g., the *creates* property of &ns2#rodin). Finally, resources can be *multiply classified* under several classes (e.g., &ns2#rodin is a **Painter** and a **Sculptor**). An RDF statement is simply an edge labeled with a property type, whereas an RDF description introduces a semistructured data graph. The semantics of edge and node labels in this graph is given by one or more associated RDF schemas.

Definition 2. *Given a population function* π, *an interpretation function is defined as follows:*

- *for a class* $c \in C$, $[\![c]\!] = \pi(c)$ *(note that* $[\![Resource]\!] = O$*),*
- *for a property type* $p \in P$, $[\![p]\!] = \{< v_1, v_2 > \mid v_1 \in [\![domain(p)]\!], v_2 \in [\![range(p)]\!]\} \cup \bigcup_{p' \prec p}[\![p']\!]$,
- *for a container type* $[\![Bag|Seq|Alt]\!] = \{1 : v_1, \ldots n : v_n\}$ *where* $v_1, \ldots v_n$ *are values in* O.

Definition 3. *An RDF description, instance of a schema RS, is a directed labeled graph* $RD = (V_D, E_D, \psi, \nu, \tau, O \cup L)$, *where:* V_D *is a set of nodes and* E_D *is a set of edges in an RDF data graph,* ψ *is the incidence function* $\psi : E_D \to V_D \times V_D$, ν *is a value function* $\nu : V_D \to O \cup L$ *and* τ *is a labeling function* $\tau : V_D \cup E_D \to N$ *which satisfies the following :*

- *for each node v in V_D, $\tau(v)$ is a set of names $n \in C \cup \{Literal, Bag, Seq, Alt\}$ where $\nu(v) \in [\![n]\!]$;*

- *for each edge ϵ from a node v to a node v' in E_D, $\tau(\epsilon)$ is a property type name $p \in P \cup \{1, 2, \ldots\}$, such that $\nu(v) \in [\![domain(p)]\!]$ and $\nu(v') \in [\![range(p)]\!]$; additionally, if $p \in \{1, 2, \ldots\}$, v should be of a container type: $(Bag|Seq|Alt) \in \tau(v)$.*

It should be stressed that our RDF graph model roughly corresponds to a finite, many-sorted relational structure. In fact, besides literal values and resource identifiers, the model relies on relations for class or property extents and containers. Note that resource URIs and names of class or property types may also be considered as values (i.e., strings), denoted as **val**. Then, an RDF data graph can be viewed as an instance of the following schema (with unnamed tuples):

$$\texttt{cls(val)} \quad \texttt{prop(val, val)} \quad \texttt{cont(val, val, val)}$$

Here **cls**, **prop** and **cont** correspond to specific schema classes, property types and to the *Bag, Seq, Alt* container types, respectively. Then $\texttt{prop}(r1,r2)$ indicates that $r1,r2$ are resource URIs connected through an edge labeled **prop**, while $\texttt{cont}(s1,1,r2)$ indicates that the first member of container value $s1$ is the resource $r2$. RDF schema vocabularies can also be represented using the relations **Class** and **Property** as well as two additional relations capturing the partial ordering ($\prec$) of classes and property types.

3   The RQL Query Language

In this section, we present the language *RQL* which allows us to query semi-structured RDF descriptions using taxonomies of node and edge labels defined in an RDF schema. The following examples depict the use of *generalized path expressions* with variables on both kinds of labels.

Q1: *Find the resources that are classified as both, Painter and Sculptor.*

```
select X
from    X Painter, Y Sculptor
where   X = Y
```

X
&rodin

Q1 is a simple, OQL-like query, with two variables ranging over sets of nodes. One of the original features of *RQL* is the ability to also consider property-types as entry-points to a semistructured RDF (meta)data graph. **Q2** depicts this functionality.

Q2: *Find the resources that "created" something, and their creations*

```
select X, Y
from    {X}creates{Y}
```

source	target
&rodin	&crucifixion
&picasso	&guernica
&picasso	&womanbird
&claudel	&eternalidol

In **Q2**, the variables X and Y are range restricted to the *source* and *target* (considered as position indices) values of the **creates** extend (including instances of the *sub-properties* of **creates**). We actually treat a property-type as a binary relationship over its domain and range, whose interpretation is a set of ordered tuples. Using these basic constructs, we can now introduce queries on node and edge labels.

Q3: *Find the resources created by a Painter, which have material "oil on canvas".*

```
select Y
from    {X:$C}creates{Y}.has_material{Z}
where   $C = Painter, Z = "oil on canvas"
```

Y
&guernica
&woman

Q3 essentially implies a navigation through the structure of descriptions and a filtering on both RDF data and schema information. Data variables, like Y and Z are range-restricted to the *target* and *source* values respectively of the *creates* and *has_material* extents. Schema variables, prefixed with the symbol $, are range restricted to the meta-collections **Class** and **Property**. In **Q3**, C denotes a class name variable, which is valuated to the domain (e.g., **Artist**) of the property *creates* and its subclasses (e.g., **Painter** and **Sculptor**). Then, the first condition in the **where** clause restricts C to **Painter**. The expression "$X : C" (similar to a cast) restricts the *source* values of the *creates* extent only to the **Painter** instances, as for example, **&ns2#rodin** and **&ns2#picasso**. Note that if the class name in the **where** clause is not a valid subclass of the domain of *creates*, the query will return an empty answer. Moreover, the composition of paths, through the "." operator in the **from** clause, implies a join between the extents of *creates* and *has_material* on their *target* and *source* values respectively. This way, RQL captures the existential semantics of navigation in semistructured data graphs: there exist two "*paints*" properties for **&ns2#picasso** while there is no "*has_material*" property for **&ns2#crucifix**, created by **&ns2#rodin** (declared also as a **Painter**). More formally, **Q3** is interpreted as:

$$\{v_2|\ c_1 \in C, c_1 \prec domain(creates), v_1 \in [\![c_1]\!], < v_1, v_2 > \in [\![creates]\!],$$
$$< v_2, v_3 > \in [\![has\_material]\!],\ c_1 = Painter\ and\ v_3 = "oil\ on\ canvas"\}$$

RQL can also be used to query RDF schemas, regardless of any underlying instances. The main motivation for this is the use of RQL as a high-level language to implement schema browsing. This is justified by several reasons: a) in real applications RDF schemas may be very large, and therefore they cannot be manipulated in main memory [5]; b) due to class refinement, RDF schemas carry information about the labels of nodes and edges which is only implicitly stated in the schema graph (e.g., by inheritance of properties). Consider, for instance, the following query computing all the outgoing edges of a specific node (or nodes) in the schema graph:

Q4: *Find all the property types and their corresponding range, which can be used on a resource of type Painter or any of its subclasses.*

```
select $P, $Y
from    {$X}$P{$Y}
where   $X <= Painter
```

$P	$Y
creates	Artifact
creates	Painting
paints	Painting

The formal interpretation of **Q4** is:

$\{< p, c_2 > | \exists p \in P, c_1, c_2 \in C, c_1 \prec domain(p), c_2 \prec range(p), c_1 \prec Painter\}$

Some of these edges are explicitly declared in the schema (e.g. *paints*) while others are inferred from the class hierarchy (e.g. *creates*). The same is true for the target nodes of the retrieved properties (e.g., `Painting` and `Artifact`). It should be stressed that due to multiple classification of nodes (e.g., `&ns2#rodin`), we can query paths in a data graph (e.g., in **Q3**) that are not included in the result of the corresponding schema queries (e.g., **Q4**). Still, the ability of *RQL* GPEs to combine filtering conditions on both graph data and schema, permits the querying of properties emanating from resources only, according to a specific class hierarchy (e.g., view the properties of `&ns2#rodin` only as a `Painter` and not as a `Sculptor`). As a last example, we illustrate how *RQL* can be used to express the *AboutEachPrefix* retrieval function of RDF [21], returning both schema and data information.

Q5: *Tell me everything you know about the resources of the site "www.museum.gr".*

```
select X, $Z, $P, Y, $W
from   {X:$Z}$P{Y:$W}
where  Y like
       "*www.museum.gr*"
       or  X like
       "*www.museum.gr*"
```

X	$Z	$P	Y	$W
&rodin	Painter	creates	&crucifix	Painting
&rodin	Sculptor	creates	&crucifix	Painting
&picasso	Painter	paints	&guernica	Painting
&claudel	Sculptor	sculpts	&eternal	Sculpture
&picasso	Painter	fname	"Pablo"	Literal
&claudel	Sculptor	lname	"Claudel"	Literal
&guernica	Painting	hasstyle	&cubism	Style
...	...	...	...	...

Q5 will iterate over all property names (P), then for each property over its domain (Z) and range (W) classes and finally over the corresponding extents (X, Y). Finally, the result of *RQL* queries represented in this section in a tabular form (e.g., as ¬1NF relations) can be naturally captured by RDF Bag containers permitting heterogeneous member sorts (e.g., literals, URIs, sequences). Closure of *RQL* queries is ensured by supporting access operators for containers [18].

4   The RDF Metadata Management System

The metadata management system currently under development (see Figure 4) comprises three main components: the RDF validator and loader (**VRP**), the RDF description database (**DBMS**) and the query language interpreter (**RQL**).

4.1   Parsing, Validation, and Storage

The *Validating RDF Parser* (VRP) is a tool for analyzing, validating and processing RDF descriptions. Unlike existing RDF parsers (e.g. SiRPAC[2]), VRP[3] is based on standard compiler generator tools for Java, namely CUP/JFlex (similar to YACC/LEX). The stream-based parsing support of JFlex and the quick LALR grammar parsing of CUP ensure a good performance, when processing large volumes of RDF descriptions. The most distinctive feature of VRP is its ability to validate RDF descriptions against one or more schemas, as well as the schemas themselves.

RDF_Resource@4487	
URI	ns2#Picasso
rdf:type	ns1#Painter

RDF_Class@4455	
URI	ns1#Painter
rdf:type	rdfs:Class
rdfs:subClassOf	ns1#Artist

RDF_Property@5678	
URI	ns1#paints
rdf:type	rdf:Property
rdfs:subPropertyOf	ns1#creates
rdfs:domain	ns1#Painter
rdfs:range	ns1#Painting
link_list	(ns2#Picasso, ns2#Guernica)
	(ns2#Picasso, ns2#Woman)

Fig. 2. Example objects in the VRP internal model

The VRP validation module relies on an internal object model implemented in Java, separating RDF schemas from their instances. Instances of those schemas adhere to the graph model presented in section 2. More precisely, the VRP model consists of the following classes (see Figure 3): Resource, RDF_Resource, RDF_Class, RDF_Property, RDF_Container and RDF_Statement. Since, for RDF, everything is a resource, Resource is the root of the class hierarchy of the VRP internal model. Proper instances of this class represent the various resources (e.g., Web pages) in RDF descriptions which are identified by a URI (a hash map is used to transform string URIs to Java object ids). RDF_Resource is a direct subclass of Resource, representing resources with defined RDF/S properties (e.g., rdf:type, rdfs:label, rdfs:seeAlso). The other classes, RDF_Class, RDF_Property, RDF_Container and RDF_Statement,[4] are subclasses of RDF_Resource. The Java objects representing schema resources are instances of the classes RDF_Class and RDF_Property. Figure 2 shows the objects created for the resources *ns2#Picasso*, *ns1#Painter* and *ns1#paints*, from the example of Figure 1.

---

[2] http://www.w3.org/RDF/Implementations/SiRPAC/

[3] http://www.ics.forth.gr/proj/isst/RDFhttp://www.ics.forth.gr/proj/isst/RDF

[4] The RDF_Statement class represents reified statements.

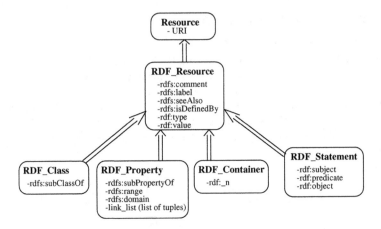

Fig. 3. VRP internal object model

This representation scheme, compared to the flat representation of triples produced by other RDF parsers, simplifies the manipulation of RDF metadata and schemas to a great extent. Firstly, the classification of resources in hierarchies makes semantics explicit. Moreover, the necessary information for loading such descriptions into a DB is straightforwardly represented in this model. Finally, by separating RDF Schemas from their instances, it allows easier manipulation of schema information, while verification of schema constraints can be performed more efficiently. This separation also facilitates a two-phase loading of schemas and their instances, as described below.

The Loader module APIs are based on the VRP internal model and comprise a number of primitive methods, which can be implemented for various DBMS technologies (e.g., relational, object). These primitive methods are defined as member functions of the classes of the VRP model, for storing the attribute values of the created objects. For example, the method `storetype()` is defined for the class `RDF_Resource`, in order to store type information of the objects. The primitive methods of each class are incorporated in a storage method defined in the respective class invoked during the loading process. The Loader takes advantage of the Java method-overriding mechanism, in order to store both RDF descriptions and schemas in a DBMS using a two-phase algorithm: During the first phase, RDF schema information (i.e., class and property descriptions) is loaded in the database, to create the corresponding storage schema. It should be stressed that the storage schema is a direct image of the associated RDFS schemas as presented in section 2. During the second phase, this schema is used to populate the database with resource descriptions. For example, Figure 4 illustrates the representation of RDF descriptions in a *relational DBMS*, using specific schema information. We should note there is significant current interest in storing semi-structured data (especially XML data) in RDBMS (e.g., [14]). Our representation consists of four tables capturing the class and property-type

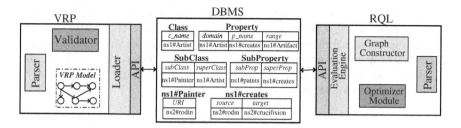

Fig. 4. System architecture

hierarchies defined in an RDF schema, namely `Class`, `Property`, `SubClass` and `SubProperty`. Then, for every new class or property loaded in the database, we create a new table to store its instances. This implementation conforms to our graph model and permits a uniform representation of both RDF descriptions and schemas, while capturing in a precise way the semantics of the latter.

4.2 Query Processing

The *RQL interpreter* consists of (a) the parser, analyzing the syntax of queries; (b) the graph constructor, reflecting the semantics of queries and (c) the evaluation engine, accessing RDF descriptions and schema information from the underlying database. As in the case of the loader, the RQL evaluation engine relies on high-level APIs that can be implemented as front-end access functions for various DBMS technologies. The development of the *RQL optimizer* is ongoing and will be mainly based on heuristic methods for query rewriting (join reordering, etc.), making use of realistic assumptions about the queried extents and exploiting possible index structures. In particular, we plan to implement indices for RDF schema classes (or property-type) hierarchies (see *Subclass* and *SubProperty* relations) in order to handle efficiently recursive access to all subclasses (or subproperties) of a given class (or property).

In applications where the RDF schema contains deep and voluminous classification hierarchies, queries accessing subclasses or subproperties of a given class or property respectively, are extremely time consuming. As demonstrated in [5], in cultural applications a schema could consist of rather deep and broad taxonomies of concepts (*terms*) originating from application specific vocabularies. In [5] the authors demonstrate the creation of an *RDF* schema by integrating the rather shallow *ICOM/CIDOC Reference Model* [16] and the rich *Art & Architecture Thesaurus* [1]. The former is a conceptual schema defined by the International Council of Museums to describe cultural information, containing around 30 concepts and 60 roles. The latter is one of the largest thesauri in the area of western art and historical terminology containing around 28.000 terms. In the schema resulting from the integration of the above conceptual structures, ICOM/CIDOC concepts and AAT terms are modeled as *RDF classes*, the latter considered as *sub-classes* of the former, organised in monohierarchical inheritance taxonomies. Those simple inheritance hierarchies are rather deep and broad and queries that require access to the subtree of a given class or property are essentially traversal queries over the *SubClass* relation of Figure 4, and are rather costly. The idea is to transform such traversal queries into interval queries on a linear domain, that can be answered efficiently by standard DBMS index structures. To do this, noder names are replaced by *ids* for which a convenient total order exists. An encoding to provide those *ids* is exposed in detail in [5].

5   Conclusions and Future Work

This paper puts forth the idea that declarative query languages for metadata, like *RQL*, open new perspectives in the effective and efficient support of WWW applications. *RQL* can be used as high-level language to access various RDF metadata repositories, by exploiting its ability to uniformly query (meta)data and schema vocabularies and to handle incomplete information. *RQL* can exploit transparently taxonomies of classes in order to facilitate querying of complex semistructured data using only few abstract labels. The paper also presents an architecture for metadata management comprising efficient mechanisms for parsing and validating RDF descriptions, loading into a DBMS and *RQL* query processing and optimization.

Current research and development efforts focus on desining appropriate access path selection mechanisms and heuristic methods for query rewriting and optimization. Appropriate index structures for reducing the cost of recursive querying of deep hierarchies need to be devised as well. Specifically, an implementation of hierarchy linearization is under way, exploring alternative node encodings. The performance of the system will be assessed using benchmarks for relational and object-oriented DBMS platforms.

References

1. The Art & Architecture Thesaurus.   http://www.ahip.getty.edu/vocabulary/-aat_intro.html.
2. S. Abiteboul, P. Buneman, and D. Suciu. *Data on the Web: From Relations to Semistructured Data and XML*. Morgan Kaufmann, 1999.
3. S. Abiteboul, S. Cluet, V. Christophides, T. Milo, G. Moerkotte, and J. Siméon. Querying Documents in Object Databases. *International Journal on Digital Libraries*, 1(1):5–18, April 1997.
4. S. Abiteboul, D. Quass, J. McHugh, J. Widom, and J. Wiener. The Lorel Query Language for Semistructured Data. *International Journal on Digital Libraries*, 1(1):68–88, April 1997.
5. B. Amann and I. Fundulaki. Integrating Ontologies and Thesauri to Build RDF Schemas. In *ECDL-99: Research and Advanced Technologies for Digital Libraries*, pages 234–253, Paris, France, September 1999.
6. B. Amann and M. Scholl. GRAM: A Graph Model and Query Language. In *Proceedings of the ECHT'92 European Conference on Hypermedia Technologies*, pages 201–211. ACM Press, December 1992.

7. D. Brickley and R.V. Guha. Resource Description Framework (RDF) Schema Specification. Technical report, World Wide Web Consortium, 1999. W3C Proposed Recommendation 03 March 1999.

8. P. Buneman, S.B. Davidson, and D. Suciu. Programming Constructs for Unstructured Data. In *Proceedings of International Workshop on Database Programming Languages*, Gubbio, Italy, 1995.

9. R.G.G. Cattell and D. Barry. *The Object Database Standard ODMG 2.0*. Morgan Kaufmann, 1997.

10. S. Ceri, S. Comai, E. Damiani, P. Fraternali, S. Paraboschi, and L. Tanca. XML-GL: a Graphical Language for Querying and Restructuring XML Documents. In *Proceedings of International WWW Conference*, Toronto, Canada, 1999.

11. V. Christophides, S. Cluet, and J. Siméon. On Wrapping Query Languages and Efficient XML Integration. In *Proceedings of ACM SIGMOD*, Dallas, 2000.

12. S. Cluet, C. Delobel, J. Siméon, and K. Smaga. Your Mediators Need Data Conversion! In *Proceedings of ACM SIGMOD*, pages 177–188, Seattle, 1998.

13. L. Dempsey and R. Heery. DESIRE: Development of a European Service for Information on Research and Education, 1997. http://www.ukoln.ac.uk/metadata/-desire/overview/rev_ti.htm.

14. A. Deutsch, M. Fernandez, and D. Suciu. Storing Semistructured Data with STORED. In *Proceedings of ACM SIGMOD*, pages 431–442, Philadelphia, 1999.

15. A. Deutsch, M.F. Fernandez, D. Florescu, A. Levy, and D. Suciu. A Query Language for XML. In *Proceedings of the 8th International WWW Conference*, Toronto, 1999.

16. M. Doerr and Nick Crofts. Electronic Communication on Diverse Data - The Role of an Object-Oriented CIDOC Reference Model. In *CIDOC'98 Conference*, Melbourne, Australia, October 1998.

17. M.F. Fernandez, D. Florescu, J. Kang, A.Y. Levy, and D. Suciu. System Demonstration - Strudel: A Web-site Management System. In *Proceedings of ACM SIGMOD*, Tucson, AZ., May 1997. Exhibition Program.

18. G. Karvounarakis, V. Christophides, and D. Plexousakis. Querying Semistructured (Meta)data and Schemas on the Web: The case of RDF & RDFS. Technical Report 269, ICS-FORTH, 2000. Available at: http://www.ics.forth.gr/proj/isst/RDF/rdfquerying.pdf.

19. M. Kifer, W. Kim, and Y. Sagiv. Querying object-oriented databases. In *Proceedings of the ACM SIGMOD*, pages 393–402, 1992.

20. L.V.S. Lakshmanan, F. Sadri, and I.N. Subramanian. SchemaSQL - a language for interoperability in relational multi-database systems. In *Proceedings of International Conference on Very Large Databases (VLDB)*, pages 239–250, Bombay, India, September 1996.

21. O. Lassila and R. Swick. Resource Description Framework (RDF) Model and Syntax Specification. Technical report, World Wide Web Consortium, 1999. W3C Recommendation 22 February 1999.

22. I.S. Mumick and K.A. Ross. Noodle: A Language for Declarative Querying in an Object-Oriented Database. In *Proceedings of International Conference on Deductive and Object-Oriented Databases (DOOD)*, pages 360–378, December 1993.

23. Y. Papakonstantinou, H. Garcia-Molina, and J. Widom. Object Exchange Across Heterogeneous Information Sources. In *Proceedings of IEEE International Conference on Data Engineering (ICDE)*, pages 251–260, Taipei, Taiwan, March 1995.

An Example-Based Environment for Wrapper Generation*

Paulo B. Golgher, Alberto H. F. Laender, Altigran S. da Silva**, and
Berthier Ribeiro-Neto

Departament of Computer Science
Federal University of Minas Gerais
31270-010 Belo Horizonte MG Brazil
{golgher,laender,alti,berthier}@dcc.ufmg.br

Abstract. In the so-called Web information systems, the role of extracting data of interest from Web sites is played by software components generically known as wrappers. As a result, the existence of flexible tools for designing, developing and maintaining wrappers is crucial. In this paper, we present WByE (Wrapping By Example), a user-oriented set of tools for helping the user to build wrappers. WByE is based on information implicitly provided by the user by means of suitable and intuitive interfaces. It includes two components: the ASByE tool, used for generating specifications on how to fetch desired pages (be them static or dynamic), and the DEByE tool, used for the extraction of data implicitly present in the fetched pages.

1 Introduction

One of the most important features of the so-called *Web information systems* is the capability of incorporating data from many different Web sites. In such systems, the role of extracting data of interest from an specific Web site is played by software components generically known as *wrappers*. As pointed out in [16], the task performed by a wrapper roughly involves three steps: (1) the fetching of the pages from the Web site; (2) the identification and extraction of data (objects) implicitly present in the fetched pages; and (3) the storage of the extracted data in suitable a format (e.g., XML, relational tables, etc.) for further manipulation. The range of Web applications requiring wrappers is enormous and, therefore, the existence of flexible tools for designing, developing, and maintaining them is crucial.

Motivated by this, many works have been developed in the last few years that address the problem of wrapper generation. The pioneer works were carried out using very sophisticated mechanisms, but which required some expertise from the user in subjects such as automata, grammars, agents, internet protocols,

---

* This work is supported by Project SIAM (grant MCT/FINEP/PRONEX 76.97.1016.00) and by individual research grants from CNPq and CAPES.
** On leave from the University of Amazonas, Brazil.

S.W. Liddle, H.C. Mayr, B. Thalheim (Eds.): ER 2000 Workshop, LNCS 1921, pp. 152–164, 2000.

and computer programming in general. This is the case of many very successful projects such as *ARANEUS* [3], *TSIMMIS* [6], *LORE* [14], *ARIADNE* [13], *FLORID* [12], and *W3QS* [8] to name just a few.

More recently, new approaches for wrapper generation have emerged that are oriented towards the less experienced Web user with few or none programming skills. This is the case for tools such as *NoDoSe* [2], *W4F* [16], and *XWRAP* [11]. Most of these approaches, however, address the problem of fetching Web pages in a simplistic manner or assume that the user already has collected the documents in a separate process.

In this paper, we present the *WByE (Wrapping By Example)* environment, a user-oriented set of tools for helping the user in the task of fetching and extracting data from Web sites. The ground for the development of such an environment is the idea that the user is able to interact with high-level graphical interfaces to provide examples of how steps (1) and (2) mentioned before should be performed for obtaining data of interest from a particular Web site. These examples are then used to automatically generate low-level specifications (including page fetching commands and data extraction patterns) that will be used to automatically collect and extract data from the Web site.

The *WByE* environment is composed of two integrated tools named *ASByE* (Agent Specification By Example) and *DEByE* (Data Extraction By Example). The first tool is used to generate a *page fetching plan* (PFP) that guides the behavior of an agent responsible for fetching a set of pages from the target Web site. The user interacts with the tool's interface to provide examples of how to reach the desired pages within a site, to fill any forms needed, and to navigate a set of related pages to form a collection. The generated fetching plan specifies how the agent will perform these tasks automatically when invoked. ASByE is specially suitable for building plans to fetch pages generated automatically as a result of filling an HTML form.

The second tool, DEByE, is used to specify how to extract data from the pages fetched and to logically organize them according to the user's perception of the implicit structure of the data in the pages. The DEByE's interface is based on a metaphor of nested tables which are built by the user by cutting and pasting pieces of data present on a sample page of the Web site he/she is interested in. From the examples provided through the assembled tables, the tool derives an *object extraction pattern (OEP)* which describes the textual context and the structure of the objects to be extracted.

Once generated by the corresponding tools, the PFP and the OEP are given as input to a general purpose wrapper, which includes a page fetching component and an extractor component. The wrapper then automatically performs the page fetching and the data extraction tasks, and stores the extracted data according to an XML-based format. if the wrapper needs to be modified due to changes in the site, the WByE environment allows the user to revise the PFP and/or the OEP accordingly. Fig. 1 illustrates the general framework provided by the WByE environment.

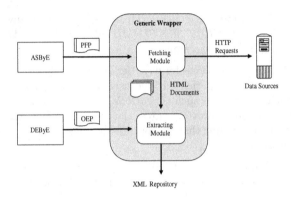

Fig. 1. Wrapper generation using WByE.

The DEByE tool and the DEByE approach for data extraction have been detailed described in [9,15]. For this reason, in the present paper, we focus our discussion on the features and the functionality of the ASByE tool. Thus, the rest of the paper is organized as follows. Section 2 describes the ASByE tool. Section 3 presents an example of a wrapper constructed for a popular Web site using the WByE environment. Finally, Section 4 concludes the paper.

2   The ASByE Tool

In this section, we give an overview of the ASByE tool. First, we present the visual paradigm used by its user interface. Next, we discuss the features related to the fetching of sets of static pages stored in a Web server. Finally, we show how the tool addresses the problem of collecting sets of dynamic pages, i.e, pages dynamically generated as a result of filling HTTP forms.

2.1   Visual Web Exploration

The user interface of the ASByE tool uses a graph-like structure to represent a portion of the Web. The nodes displayed in its work space represent pages and directed arcs represent hyperlinks. Similar metaphors were previously used by the *Hy+* system [7] and by the diagrams of the *ARANEUS* data model [3], which inspired the "look and feel" of the interface. The user navigates from node to node exploring the hyperlinks according to his/her interests. Fig. 2 shows a typical usage session with the interface, while Fig. 3 shows the types of arcs and nodes used by the interface to represent the different pages a user finds when browsing a Web site. The source nodes in the graph (i.e., the ones not pointed by any other node) are called *Web entry points* and are directly selected by the user using a dialog box where he/she enters the URL of the page from where he/she wants to start the exploration. The tool then fetches the page and builds a node corresponding to it. From this point onwards, the user can select for

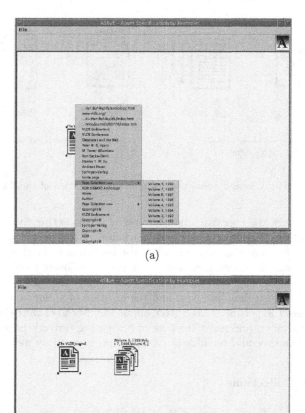

(a)

(b)

Fig. 2. Usage session for Web Exploration.

each node an operation he/she wants to perform. The set of operations available depends on the type of the node reached. The most common and simple operation allows the user to select a hyperlink to explore, what is done by selecting one of the hyperlinks shown in a pop-up menu. When a hyperlink is selected, the corresponding page is fetched and a new node is created in the interface's work space to represent it.

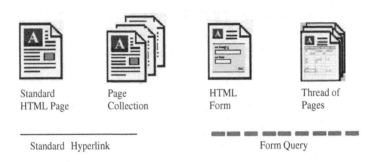

Fig. 3. Arcs and nodes used to represent a portion of the Web.

Alternatively, the user can follow a hyperlink by selecting the "View in Browser" operation. This operation opens a window of a browser[1] where the user can analyze the page contents and select the hyperlink of his/her interest. Due to specific instrumentation included in the page previously to its presentation, the ASByE tool is notified of the hyperlink selected by the user, what causes the fetching of the corresponding page and the creation of a new node to represent the newly select hyperlink. This feature allows the user to explore the Web in the traditional manner, and eases the task of navigating through pages with many hyperlinks, which would be difficult to represent in a pop-up menu.

2.2 Page Collections

In many cases, the user is interested in treating a set of logically related pages as a collection. As an example, suppose one is interested in collecting from the DB&LP Web site [4] the pages that contain information on all the volumes of the VLDB Journal. These pages can be reached from the VLDB Journal Page on that site (see Fig. 4). It will be tedious or even unfeasible for the user to specify that he/she wants to individually collect each page corresponding to each volume. To cope with this, the ASByE tool tries to infer, from each page represented by an icon, sets of related links in such a way that a single collecting specification can be generated for the whole collection of pages referred by these links. The tool uses several heuristics to determine which links will be grouped. These heuristics are based solely on the position and on the labels of the hyperlinks. As an additional advantage, when more similar links are added to the page, the corresponding pages will be automatically included in the collection.

In the interface, when the user selects a "Page Collection" entry in the hyperlinks menu (see Fig. 3), a new node is created with an icon that represents a collection of pages. We note that no page is fetched at this moment and that this node is considered terminal. Moreover, the only operation available is for the generation of the fetching plan for the page collection.

---

[1] Currently, only Netscape is supported, but the tool can be easily extended to support the MS Explorer.

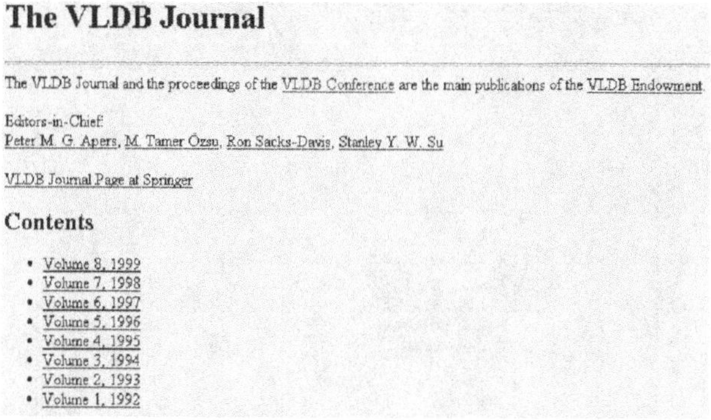

Fig. 4. A snapshot of the VLDB Journal page from the DB&LP site.

For identifying a set of links in a page as links to a page collection, the following heuristics are considered: (1) **Hyperlinks Proximity** - Sets of very near hyperlinks in a page are considered candidates to form a collection (e.g., several links separated by a single comma); (2) **Similarity in the URLs** - Similarities in the URLs refered by a group of hyperlinks count positively towards considering them as a collection (e.g., URLs formed by the same directory path); (3) **URL Hosting** - URLs belonging to different HTTP servers are usually not considered as part of a collection; (4) **Enumeration in the Hyperlink Labels** - A group of hyperlinks whose labels form an explicit enumeration is a strong candidate to be considered as a collection (Fig. 4 presents an example of such a case); (5) **Number of Hyperlinks in the Page** - The fewer is the number of hyperlinks the stronger is the support needed for the above heuristics to be taken in consideration when identifying page collections.

The heuristics adopted work well for pages where the semantic context is well defined, as is the case for the page shown in Fig. 4. However, in many other cases the heuristics may fail for pages whose subject is less focused, as it occurs in pages from the so-called "Web Portals". For cases like this, the interface allows the user to group or ungroup links by his/herself to form customized page collections.

2.3 Dealing with Forms

A recent study [10] estimates that 80% of the Web contents are located on the so-called *hidden Web*, i.e., on Web pages which are dynamically generated. In most of the cases, these pages are generated by programs invoked through HTML forms. Thus, an important feature for any Web agent is the ability of automatically filling these forms to retrieve the generated pages. Moreover, in many cases the agent must take parameters as input to supply different values to form fields each time they run. The ASByE tool provides means for the user to include in a page fetching plan instructions to fill HTTP forms, with default values or with values derived from parameters supplied as input at the moment

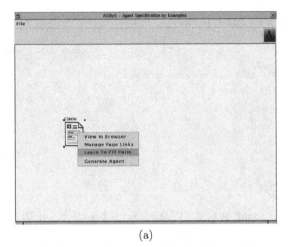

Fig. 5. Dealing with forms in the ASByE tool.

the plan is executed. Here again, the interface asks the user to provide an example of how to fill the form and uses this information to generate instructions for filling it in the collecting specification.

Fig. 5(a) shows an example of navigation that leads to a page containing a form. For this page, the user selected the "Learn to Fill Form Operation". This operation opens a window of a browser displaying a modified version of the form page. In this version, a small bullet is included near to each form field, as shown in Fig. 5(b) (notice the small round shaped A).

When the user clicks one of these bullets, a pop-up box is opened for the user to specify: (a) if the field should be a parameter or if a default value should be used all the time and (b) a nickname to be used when referring to the field. This nickname eases the task of later assigning values to the parameterized fields, since the user does not need to know the name coded in the HTML description of the page, which might not be representative of the meaning of that field. After filling the fields he/she is interested in, the user finally submits the form as he/she would normally do in an ordinary operation. The tool then uses this information to generate the correct instructions for filling and submitting the form within the page collecting specification.

In most cases, the result of submitting an HTML form is a linked thread of pages. Thus, after the sample form submission, the interface creates a new node using an icon that indicates an answer thread. Next, we discuss how we collect pages that compose an answer thread in our tool.

2.4 Collecting Threads of Answer Pages

Although there can be many ways to generate answer pages that result from filling HTML forms, almost all services available on the Web include, in each page, a hyperlink or a button to the next page in the answer. We call such a set of pages a *thread*.

To generate collecting specifications for a thread of answer pages, we again rely on a user action. By selecting the operation "Learn to Follow" from an answer thread node, the user is presented with the first page of the thread shown in the browser window. Actually, this page is a modified version of the first page of the thread, instrumented in such a way that every link selection and button pressing is trapped by the tool. Then, the user is asked to select the link or to press the button that leads to the next page in the thread. This action is trapped by the tool and its request (a GET or POST) is analyzed with the corresponding parameters. Based on this, a number of "hints" to control the behavior of the fetching agent is coded in the page fetching plan. These hints are used to determine how to fetch, from the first page of a thread, all subsequent pages until the last page has been reached.

Depending on the type of mechanism used to fetch the subsequent pages (link or button), one of two available heuristics is used and coded in the PFP. If the user selects a link during the "Learn To Follow" operation, the ASByE tool generates a pattern expression that extracts the link used to retrieve the subsequent results. The tool uses techniques adopted by the DEByE tool [15] for generating this pattern expression. While there is a match for this regular expression, the pages are collected until no more links that match the expression are found (Heuristic 1).

When a button or other mechanism causes a form to be posted for retrieving subsequent pages, the heuristic applied is a bit more complex. The heuristic is based on the fact that generally a variable is passed when the form is posted to inform the page of the thread to be fetched. For instance, the following set of

URLs below corresponds to the URLs used to fetch the second and third pages of a given query on the eBay site [5] (notice the `skip` variable):

```
http://search.ebay.com/search/search.dll?MfcISAPICommand=GetResult
&query=guitar&SortProperty=MetaEndSort&SortOrder=%5Ba%5D&skip=50

http://search.ebay.com/search/search.dll?MfcISAPICommand=GetResult
&query=guitar&SortProperty=MetaEndSort&SortOrder=%5Ba%5D&skip=100
```

Thus, the ASByE tool traps the form request generated by the user during the "Learn to Follow" operation and uses this information to infer the variables that change their values on the next request for a page of the thread. This is done by parsing the retrieved document and extracting the values of the corresponding form that was posted in the example given by the user. The numerical values that change from one posting to the other are considered the values that are responsible for determining the current page of the thread. Therefore, for fetching the remaining thread of pages, the tool generates information on how to enumerate the values of these variables (Heuristic 2).

In order to validate the two heuristics presented, we developed PFPs for 10 of the most popular sites on the Web, according to the 100Hot site [1]. Some of the most popular sites (such as the `Quote.com` site) do not provide any means for generating a thread of answer pages and were therefore discarded. Also, although the eBay and Monster sites were not the 9th and 10th of 100Hot list, they were added because there are not many sites that require the use of the Heuristic 2. The results are shown in Table 1. We observe that the results are very good, except in the case of the `Lycos` site. In this case, ASByE failed to generate a correct pattern for Heuristic 1 because the first page of the thread used to generate the pattern had a completely different layout than the other pages in the thread.

Site	Heuristic	Query	Pages Available	Pages Retrieved
Yahoo!	Heuristic 1	"dynamic page crawler"	10	10 (100%)
Microsoft	Heuristic 1	"windows"	9	9 (100%)
AOL	Heuristic 1	"page crawler"	41	41 (100%)
Lycos	Heuristic 1	"dynamic page crawler"	100	1 (1%)
Excite	Heuristic 2	"dynamic page crawler"	100	100 (100%)
Altavista	Heuristic 1	"dynamic page crawler"	20	20 (100%)
GO	Heuristic 1	"dynamic page crawler"	40	40 (100%)
Amazon	Heuristic 1	"database system"	5	5 (100%)
eBay	Heuristic 2	"guitar"	218	218 (100%)
Monster	Heuristic 2	"search engine"	8	8 (100%)

Table 1. Validation of the heuristics for thread collection.

3   An Example

In order to illustrate a typical interaction of a user with the WByE environment, in this section we present the development of a wrapper intended to fetch and extract data from the Amazon bookstore site. This wrapper could be used in many applications, such as shop comparison and mediators. The wrapper generation process includes two major steps: (1) the generation of the page fetching plan (PFP), with information on how to fetch the data, and (2) the generation of the object extraction pattern (OEP) which describe how to structure and extract the data of interest.

The first issue to address when generating the PFP is to locate the HTML page which allows the user to make queries to retrieve the data. Using one of the exploring features presented in the previous section, the user can easily reach the Amazon's advanced search page, which provides author and title based search, as illustrated in Fig. 6(a).

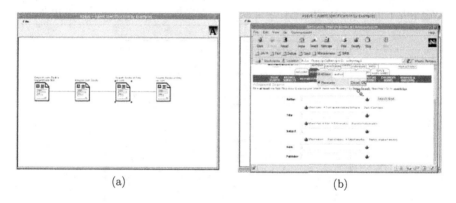

(a)   (b)

Fig. 6. Locating the Amazon's advanced search page and defining fields properties.

Then, the user determines which query fields should be parameterized and, optionally, gives a nickname for each field using the special bullets inserted in the page, as illustrated in Fig. 6(b). In this case, the author and title query fields were chosen, thus allowing the agent to fetch data related to different authors and/or titles each time it is executed. After defining these properties of each field, the user submits the form as he/she would do in a traditional navigation. The ASByE tool traps the information submitted, fetches the corresponding pages, and represents them by an icon in its interface, as shown in Fig. 7(a).

The result of submitting a query to the Amazon Bookstore site is a thread of one or more pages, connected by a "Next Results" link which retrieves the next 50 results related to the initial query. Thus, the user has to use the "Learn To Follow" operation in order to give an example of how to follow this thread of pages. This operation opens a browser window showing the first page of the thread. The user then gives the example by clicking in the "Next Results" link

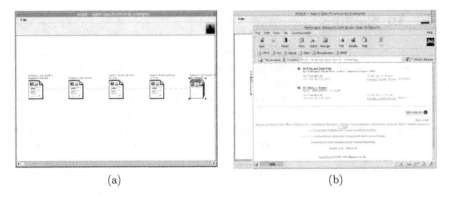

Fig. 7. Representation of the results (a) and learning to follow the Amazon's site.

of this page. Fig. 7(b) illustrates this step. Finally, the user generates the page fetching plan. An excerpt of PFP generated for this example are shown in Fig. 8.

```
<PFP sourceref='Amazon'>
    <PLAN>
            <REQUEST method=1>
                    http://www.amazon.com/exec/obidos/search-handle-form/
            </REQUEST>
            <REQUEST_FIELDS>
                    <VALUE field='index' parameter='no'> books </VALUE>
                    <VALUE field='query-0' parameter='yes'> internet </VALUE>
                    ...

            </REQUEST_FIELDS>
            <FOLLOW_METHOD>
                    <METHOD> 1 </METHOD>
                    <LINK_POS> 200 </LINK_POS>
                    <LINK_LEN> 91 </LINK_LEN>
        ...

    </PLAN>
</PFP>
```

Fig. 8. An excerpt of the generated PFP for the Amazon site.

Having generated the PFP, the next step is the generation of the OEP. Using the DEByE tool, the user loads a sample page from the set of answer pages returned by the Amazon site and gives examples of the data of interest. An OEP is then generated. Fig. 9 illustrates the excerpt of the OEP generated for the given example. The PFP and the OEP generated in the previous steps are then fed to the generic wrapper which fetches and extracts the data, storing them in an XML format [9] in a repository for further use.

4   Conclusion and Future Work

In this paper we discussed WByE, a software environment for fetching and extracting data from Web sites. Its main distinctive feature is to be fully based

```
<?xml version = "1.0"?>
  <OBJECTS>
    <TUPLE type="Book">
      <ATOM type="Title">
        <PATTERN>
          <![CDATA[<b>[\s]*?<a [^<]+>[\s]*? ...
        </PATTERN>
      </ATOM>
      <ATOM type="Authors">
        <PATTERN>
          <![CDATA[<dd>[\s]*?([\x20-\x3B\x3D\x3F-\x7E\xA0-\xFF]+?) ...
        </PATTERN>
      </ATOM>

      ...

    </TUPLE>
  </OBJECTS>
```

Fig. 9. An excerpt of the generated OEP for the Amazon site.

on information implicitly provided by the user by means of suitable and intuitive interfaces based on high level abstractions such as nested tables an directed graphs. WByE includes two components: the ASByE (Agent Specification By Example) tool, used for generating specifications on how to fetch desired pages (be them static or dynamic) from some Web site, and the DEByE (Data Extraction By Example) tool, used for generating specification on how to extract data implicitly present in the fetched pages.

Since the DEByE tool was already introduced in [9,15], we focused our discussion on the ASByE tool. Thus, we described its main features, presented the results of an experiment we have done with some popular Web sites, and discussed an example of its usage within the WByE environment.

The ASByE tool has been designed having in mind a large class of Web sites that provide huge amounts of data buried into HTML pages, be them static or dynamically generated. Observing the features found in many of these sites, we implemented the ASByE tool to deal with common situations found in them, such as page collections, forms, and page threads. As a consequence, for sites that include features based on technologies such as Java scripts and Java or Flash applets, i.e., browser-processed generic code, the ASByE tool would fail in generating a proper collecting specification.

Compared to other systems such as W3QS, FLORID, ARANEUS, XWRAP, and W4F, ASByE includes some important features that are unique. First, it requires no code writing; second, it includes a navigation metaphor based on a intuitive visual paradigm; third, it supports automatic filling of HTML forms; and fourth, it allows handling page collections and threads of answer pages. These features make ASByE specially suitable for cases when the generated wrapper has to be maintained due to changes in the sites of interest.

Although we have discussed the use of the ASByE tool in the context of wrapper generation, its range of application is very much broader. For example, we are now investigating its use for addressing a very well known problem in general purpose Web crawlers used in search engines, which is the systematic collection of dynamically generated Web pages.

164   P. Golgher et al.

References

1. 100 HOT. 100 Hot Web Site. http://www.100hot.com/.
2. ADELBERG, B. NoDoSE - A Tool for Semi-Automatically Extracting Structured and Semistructured Data from Text Documents. In *Proceedings of the ACM SIGMOD Conference on Management of Data* (Seattle, Washington, 1998), pp. 283–294.
3. ATZENI, P., MECCA, G., AND MERIALDO, P. Semistructured und Structured Data in the Web: Going Back and Forth. *SIGMOD Record 26*, 4 (1997), 16–23.
4. DB&LP. DB&LP's Index to ACM TODS. *http://www.informatik.uni-trier.de/~ley/db/journals/tods/index.html.*
5. EBAY. eBay Web Site. http://www.ebay.com/.
6. HAMMER, J., GARCIA-MOLINA, H., NESTOROV, S., YERNENI, R., BREUNIG, M., AND VASSALOS, V. Template-Based Wrappers in the TSIMMIS Experience. In *Proceedings of the ACM SIGMOD Conference on Management of Data* (Tucson, Arizona, 1997), pp. 532–535.
7. HASAN, M., MENDELZON, A., AND VISTA, D. Applying Database Visualization to the World Wide Web. *ACM SIGMOD Record 25*, 4 (1996), 40–44.
8. KONOPNICKI, D., AND SHMUELI, O. Information Gathering in the World-Wide Web: The W3QL Query Language and the W3QS System. *ACM Transactions on Database Systems (TODS) 23*, 4 (1998), 369–410.
9. LAENDER, A. H. F., RIBEIRO-NETO, B., DA SILVA, A. S., AND SILVA, E. S. Representing Web Data as Complex Objects. In *Proceedings of the First International Conference on Electronic Commerce and Web Technologies - EC-Web 2000* (Greenwich, UK, 2000), S. Madria and G. Pernull, Eds., Lecture Notes in Computer Science.
10. LAWRENCE, S., AND GILES, C. Searching the World Wide Web. *Science 280*, 4 (1998), 98–100.
11. LIU, L., PU, C., AND HAN, W. XWRAP: An XML-enabled Wrapper Construction System for Web Information Sources. In *Proceeding of the 16th International Conference on Data Engineering* (San Diego, California, 2000), pp. 611–621.
12. LUDÄSCHER, B., HIMMERÖDER, R., LAUSEN, G., MAY, W., AND SCHELEPPHORST, C. Managing Semistrucutured Data with FLORID: a Deductice Object-Oriented Approach. *Information Systems 23*, 8 (1998), 589–614.
13. MUSLEA, I., MINTON, S., AND KNOBLOCK, C. A hierarchical approach to wrapper induction. In *Proceedings of the 3rd Conference on Autonomous Agents* (Seattle,Washington, 1999), pp. 190–199.
14. QUASS, D., WIDOM, J., GOLDMAN, R., HAAS, K., LUO, Q., MCHUGH, J., NESTOROV, S., RAJARAMAN, A., RIVERO, H., ABITEBOUL, S., ULLMAN, J. D., AND WIENER, J. L. LORE: A Lightweight Object REpository for Semistructured Data. In *Proceedings of the International ACM SIGMOD Conference on Management of Data* (Montreal, Canada, 1996), p. 549.
15. RIBEIRO-NETO, B., LAENDER, A. H. F., AND DA SILVA, A. S. Extracting Semi-Structured Data Through Examples. In *Proceedings of the Eighth ACM International Conference on Information and Knowledge Management - CIKM'99* (Kansas City, Missouri, 1999), pp. 94–101.
16. SAHUGUET, A., AND AZAVANT, F. Web Ecology: Recycling HTML pages as XML documents using W4F. In *Proceedings of the Second International Workshop on the Web and Databases* (Philadelphia, Pennsylvania, 1999), pp. 31–26.

Flexible Category Structure for Supporting WWW Retrieval

Yoshiaki Takata, Kokoro Nakagawa, and Hiroyuki Seki

Graduate School of Information Science
Nara Institute of Science and Technology
{y-takata,kokoro-n,seki}@is.aist-nara.ac.jp

Abstract. A method for supporting WWW retrieval by constructing a flexible category structure adaptable to the user's search intention is proposed. The method uses categorization viewpoints as *a priori* knowledge, where a categorization viewpoint is a finite set of consistent category names. A set of documents retrieved by initial keywords is decomposed by categorization viewpoints and each decomposition is scored by clearness or entropy. The user selects an appropriate decomposition by considering the score. The decomposition is recursively performed until a category structure of reasonable size is obtained. Experimental results show that the sets of documents decomposed by the proposed method have higher precision than those decomposed by clustering (K-means). It is also shown that both the scores based on clearness and entropy of the decomposition have relatively high correlation with the precision.

1 Introduction

As the number of documents increases explosively on the World Wide Web (WWW), the necessity for an effective searching method for WWW retrieval is rapidly growing. More than a hundred Web sites (*search services*) for aiding WWW retrieval exist; however, usability problems still remain.

A search engine such as *Alta Vista* provides a user with a list of matching documents relevant to input keywords. However, the results are often confused in an abundance of unrelated documents caused by inappropriate keywords. Selecting appropriate keywords which can extract only documents relevant to the user's purpose of retrieval is difficult, especially for users unfamiliar with searching.

A directory service such as *Yahoo!* provides a user with a human-classified collection of Web-site descriptions. By traversing categories in the directory, the user can browse documents associated with the categories. This method is simpler than the one by a search engine. However, such a category structure often has too much volume to allow easy browsing. Another problem is that the categorization given by a system is not always suitable for the user's purpose of retrieval. For example, when a user is retrieving some information on official research facilities in Nara which are responsible for environmental problems, a hypothetical path in the category structure would be;

$$\text{root} \rightarrow \textit{Nara} \rightarrow \textit{Environment and Nature} \rightarrow \textit{Institutes}, \tag{1}$$

S.W. Liddle, H.C. Mayr, B. Thalheim (Eds.): ER 2000 Workshop, LNCS 1921, pp. 165–177, 2000.

but the provided category structure does not always have such a path. The relevant information is often scattered around several paths such as:

root → *Science* → *Environmental Engineering*,
root → *Regional* → *Nara* → *Health*,
root → *Society and Culture* → *Environment and Nature* → *Government Agencies*.

In the above case, the user hesitates over which category to select first, and even if the user decides a path after some consideration, she or he will only obtain a subset of documents with poor precision, since relevant documents are scattered around several categories. The problem is that the categorization criterion varies according to the user's purpose of retrieval and one fixed structure cannot cope with this versatility of criteria.

In this paper, we propose a method for supporting WWW retrieval, by providing a *small* and *flexible* category structure adaptable to the user's purpose of retrieval. It is assumed that the users are not experts on WWW retrieval, and that the purpose of retrieval is not a specialized one. A *small* category structure means a hierarchical structure that classifies a subset of documents which are related to the user's interest. Providing a small category structure solves the problem such that the structure provided by a directory service is too large to use easily. On the other hand, *flexibility* means that for each retrieval, a hierarchical structure is constructed which contains paths suitable for the user's purpose of retrieval. For the above example, where official research facilities responsible for environmental problems are retrieved, the category structure provides the path shown in (1) (cf. Figure 1 (a)). When the user is retrieving the recent topics of environmental problems, the category structure provides the following path (cf. Figure 1 (b)):

root → *News* → *Science* → *Environment* → *Ozone Depletion*.

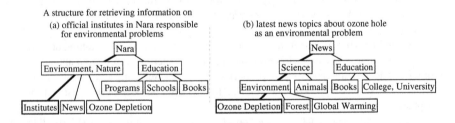

Fig. 1. Examples of Category Structures

The method uses categorization viewpoints as *a priori* knowledge, where a categorization viewpoint is a finite set of consistent category names. A set of documents retrieved by initial keywords is decomposed by categorization viewpoints and each decomposition is scored by clearness or entropy. A user selects an appropriate decomposition by considering the score. The decomposition is recursively performed until a category structure of reasonable size is obtained. As a result, the relevant documents are located in a few categories, which helps the user retrieve relevant documents. Further-more, other categories related to the user's purpose of retrieval are also located near the categories selected by the user in the category structure.

The rest of the paper is organized as follows: In Section 2, the proposed method is presented in detail. Section 3 presents the experimental results on the effectiveness of the proposed method compared with a simple keyword-based retrieval method and a traditional statistical method (K-means). The appropriateness of clearness and entropy as the score is also discussed based on the experiment.

Related Works. Many approaches to supporting WWW retrieval by estimating and adapting the user's purpose of retrieval exist. RCAAU[7] performs text data mining on the initial set of documents and provides a user with secondary keywords to obtain a set of documents with higher precision. The system in [1] also provides secondary keywords or phrases for a user. In [1], a word (called *facet*) which has high *lexical dispersion* is assumed to be useful for identifying key concepts related to the initial set of documents, and facets and noun phrases containing a facet are provided as possible subtopics related to the initial query. In [4], the user's log, such as previously input keywords, are analyzed and a query issued by the user is automatically refined to achieve higher precision and recall. These approaches are useful to drill down an initial set of retrieved documents. However, it seems difficult to decompose a document set in a consistent way and/or build a category structure according to each search intention in these methods.

Statistical approaches have been used for decomposing a set of documents. For example, [10] uses clustering to construct a category structure. Although [10] aims at constructing a single category structure for a whole set of documents, the method could be used for constructing a category structure for each retrieval if documents are restricted to those retrieved by initial keywords. A statistical approach has an advantage that *a priori* knowledge such as categorization viewpoints in this paper is not needed. On the other hand, there are some drawbacks in this approach. For a user to select a cluster appropriately, sufficient information on each cluster should be provided for the user. However, it is difficult to automatically associate a title or a summary with a cluster. In [14], a method for automatically constructing a concept hierarchy without *a priori* knowledge is proposed. Unlike clustering, each node of the concept hierarchy derived by this method is a word or a noun phrase which may simply indicate a concept. However, the method only provides a hierarchy of conceptual words, and cannot be directly used to obtain a set of documents with higher precision in itself.

HIBROWSE[11] uses portions of a thesaurus (called *views*) as *a priori* knowledge like our method. A user of HIBROWSE can see categorizations made by several views at once and can choose a keyword in any view to narrow down the temporally retrieved set. Since HIBROWSE was designed for a special purpose (searching medical biblio-graphies), the selection of views is left to a user and no method for recommending appropriate views is proposed.

Information visualization[3,12] is another approach which could be consistently incorporated into the above-mentioned methods, including ours.

2   The Method

The proposed method consists of the following two steps: (i) Identify a set of documents which roughly reflect the user's purpose of retrieval. (ii) Construct a category structure depending on the set of documents obtained in (i).

In step (ii), the system uses *categorization viewpoints*, which are built in the system as preliminary knowledge, for constructing a category structure. A categorization viewpoint is a finite set of categories. For example, a categorization viewpoint *Regional* may include *Kyoto*, *Osaka*, and *Nara*, and a viewpoint *Entertainment* may include *Books*, *Games*, *TV*, and *Travel*.

Let

$$S = \{S_j \mid 1 \le j \le m\}$$

be a set of categorization viewpoints where a categorization viewpoint $S_j = (l_j, W_j)$ is a pair of the viewpoint name l_j and a finite set of category names $W_j = \{w_{j1}, \ldots, w_{jk_j}\}$. Each W_j is assumed to be a subset of the set W of all words described in subsection 2.1.

A categorization viewpoint corresponds to a set of brother nodes (the set of children of a node) in the category structure of a directory service (Figure 1). As described in the following, different category structures are constructed depending on which categorization viewpoints are used. The system and a user construct interactively an appropriate category structure by selecting appropriate categorization viewpoints.

2.1   Representation of Documents

The *vector space model*[13] is used for defining the similarity coefficient between a document and a category. In this model, each document is represented by a vector which consists of the relevance values of each word to the document. Let

$$W = \{w_1, w_2, \ldots, w_n\}$$

be the set of all words. The feature vector of a document d is

$$c(d) = (c_{w_1,d}, c_{w_2,d}, \ldots, c_{w_n,d}),$$

where $c_{w_i,d}$ is the relevance value of word w_i to document d. The relevance value $c_{w_i,d}$ is defined as the frequency of w_i in d. In a more sophisticated model, a larger weight is given to a word which is less frequent in the document collection. Also the word frequency is normalized against the document length[2].

The *similarity coefficient* $sim(u, v)$ between two vectors u and v is defined as the normalized inner product between u and v; that is,

$$sim(u, v) = \frac{u \cdot v}{|u|\,|v|}$$

where $|u|$ represents the length of vector u.

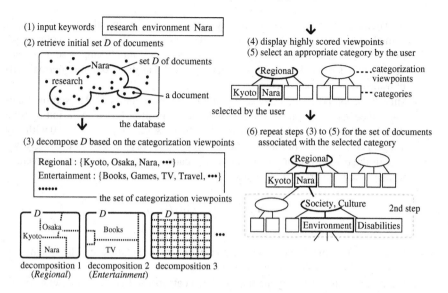

Fig. 2. Behavior of the System

2.2 The System Behavior

A user and the system perform the following steps (in Figure 2):

(1) The user inputs a few keywords to the system. Since these keywords are used for roughly limiting the documents to search, the user does not have to select them carefully.

(2) The system retrieves a set of documents which seem to relate to the input keywords from the document database. In the experimental system, documents are retrieved by the disjunctive query of the keywords. The retrieved set of documents is denoted as D.

(3) The system decomposes D into subsets by each categorization viewpoint S_j in the system; for each document $d \in D$, the system decides a category $w_j^{(d)}$ to which d belongs among the categories $W_j = \{w_{j1}, \ldots, w_{jk_j}\}$ of viewpoint S_j. The category $w_j^{(d)}$ is one of the categories in S_j which has the maximum similarity coefficient with document d. The decomposition $(D_{j1}, \ldots, D_{jk_j})$ of a set D of documents by S_j is defined as follows:

$$D_{ji} = \{d \in D \mid sim(c(w_{ji}), c(d))$$
$$= \max_{1 \leq h \leq k_j} sim(c(w_{jh}), c(d))\}.^1$$

D_{ji} is called the set of documents associated with category w_{ji}.

---

[1] For simplicity, if there are more than one h which maximizes $sim(c(w_{jh}), c(d))$, then let d belong to one of D_{jh}.

The vector $c(w_{ji})$ is called the *category vector* of w_{ji}. The initial value of $c(w_{ji})$ is defined as $(\delta_1, \delta_2, \ldots, \delta_n)$ where $\delta_k = 1$ if $w_{ji} = w_k \in W$ and $\delta_k = 0$ otherwise. Since $sim(c(w_{ji}), c(d)) = c_{w_{ji}, d}/|c(d)|$, the decomposition using these initial values is equivalent to identifying the category which has the maximum (normalized) relevance value to each document.

After the decomposition, the system scores each viewpoint for D. The scoring function will be defined in subsection 2.3. In the decomposition based on the initial value of $c(w_{ji})$, there exist many documents d such that $sim(c(w_{ji}), c(d)) = 0$ with any category w_{ji}, because of the discreteness of the category vectors. Accordingly, we smooth the categorization by redefining the category vector $c(w_{ji})$ as the average of the feature vectors of documents associated with the category w_{ji}:

$$c(w_{ji}) = \sum_{d \in D_{ji}} c(d)/|D_{ji}|.$$

(4) The system lists the viewpoints which have higher score to the user. The system also shows the categories of the listed viewpoints.
(5) The user selects a category among the listed categories. Let D' be the set of documents associated with the selected category. The user also chooses to accept D' as the final result of the session or to further decompose D' recursively into subsets.
(6) When the user chooses to decompose D' in step (5), go to step (3) with $D = D'$.

2.3   Scoring Categorization Viewpoints

We use two criteria for scoring a categorization viewpoint: the *clearness* of categorization and the *entropy* of categorization. The *score* $e_D(S_j)$ of a categorization viewpoint S_j for a set D of documents is defined as expression (2) when the clearness is used, and is defined as expression (3) when the entropy is used. The effectiveness of the two criteria is empirically compared with each other in section 3.

Clearness of categorization. We say a categorization of a set D of documents is clear when, for each document $d \in D$, we can clearly decide a category $w_j^{(d)}$ to which d belongs; that is, the similarity coefficient between d and $w_j^{(d)}$ is much larger than the similarity coefficient between d and a category other than $w_j^{(d)}$ (as Figure 3 (a)). Although a categorization viewpoint resulting in high clearness does not always reflect the user's purpose of retrieval, the viewpoint is at least informative to the user since the set of documents associated with each category is clearly separated with each other (Figure 3). The clearness of a categorization viewpoint S_j for a set D of documents is defined as

$$\sum_{d \in D} sim(c(w_j^{(d)}), c(d))/|D|, \qquad (2)$$

which is the average of the similarity coefficient between each document and the category to which the document belongs.

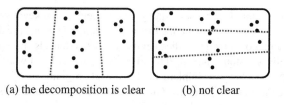

(a) the decomposition is clear   (b) not clear

Fig. 3. Clearness of Categorization

Entropy of categorization. Let $|D|$ denote the number of documents in a set D of documents. For a set D of documents and a categorization viewpoint S_j, let us define the entropy of S_j for D as

$$H(S_j) = -\sum_{i=1}^{k_j} P_i \log P_i, \tag{3}$$

where

k_j : the number of categories in S_j,

$P_i = \dfrac{|D_{ji}|}{|D|}$,

D_{ji} : the set of documents associated with category w_{ji}.

A categorization viewpoint resulting in high entropy offers an efficient means of narrowing down the documents. Assume that a set D of documents should be recursively decomposed to obtain a subset D' of D such that $|D'| < c$ for a constant c. The average number of the iterations to obtain D' (steps (3) to (6) in subsection 2.2) is smaller if a categorization viewpoint with larger entropy is selected (cf. Figure 4).

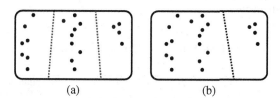

(a)   (b)

the entropy of (a) > the entropy of (b)

Fig. 4. Entropy of Categorization

3 Experiments

3.1 Overview

Our prototypic system consists of two components: the database constructor and the query processor. The database constructor preprocesses the document database, and the query processor responds to a user's query. The system uses a keyword-based text retrieval system Freya[6] for constructing the *inverted index*[5] which is a data structure to retrieve documents containing specified keywords. The system also uses a Japanese morphological analysis system ChaSen[9] for identifying the set W of words (in subsection 2.1). The query processor was implemented as a CGI program.

To evaluate the method, we defined the following four experimental systems and we conducted the two experiments described below.

- System A: this system uses clearness for scoring categorization viewpoints.
- System B: this system uses entropy for scoring categorization viewpoints.
- System C: this system does not use categorization viewpoints but decomposes a set of documents by clustering.
- System D: the keyword-based retrieval system Freya.

System C is for comparing our method which uses *a priori* knowledge with statistical methods for decomposing a set of documents. We consider K-means method[15], which is one of the simplest but most useful methods. In the experiments, we let the number of clusters $= 10$. If the centers of the clusters do not converge after 100 iterations, then terminate the iteration and exit.

We did not use a human subject but assumed an ideal user who always selects a categorization viewpoint with the highest score and categories with which a set of documents with high precision are associated. The details of the experiments will be described in the following subsections 3.2 and 3.3.

Ten topics out of the BMIR-J2 test collection[8] were used in the experiments, which are shown in Table 1. BMIR-J2 is a test collection (which includes 5,080 articles) for evaluation of information retrieval systems, based on the Mainichi Shimbun Newspaper CD-ROM '94 data collection. BMIR-J2 was constructed by the SIG Database Systems of the Information Processing Society of Japan, in collaboration with the Real World Computing Partnership.

Table 1. Topics Used in the Experiments

Name of topic [*1]	The number of relevant documents
Agricultural Chemicals	26
Beverage	60
Tax Reduction	302
Nuclear Weapon	98
The Educational Industry	17
Trend of Stock Prices	37
The Discount Brokers	36
Motion Picture	21
Export from Southeast Asia to Japan	59
Civil Harassment against Women in Employment	80

[*1]: Every topic is originally written in Japanese.

We prepared a set of categorization viewpoints (including 68 categorization viewpoints and 1,723 categories) by referring to several category structures served by existent directory services.

3.2   Experiment 1

This is a basic experiment to compare precision-recall among Systems A, B, C and D. The following procedure is performed to evaluate System A.

- Parameter: Th (> 0); A chosen category is decomposed until $|D_{jh}| \leq Th$ where D_{jh} is the set of documents associated with the category. In the experiment, we let $Th = 30$.

For each topic q in the test set (Table 1) and for each $r = 0.1, 0.2, \ldots$, and 1.0, perform the following (1) to (4).

(1) Let $Result = \emptyset$. Retrieve the set D of documents by a few keywords for topic q. Calculate the score (clearness) of each categorization viewpoint for D.
(2) Choose S_j with the highest score among the categorization viewpoints $S_1, S_2, \ldots, S_m$ ($m = 68$). Let $W_j = \{w_{j1}, w_{j2}, \ldots, w_{jk_j}\}$ be the set of categories of S_j and let D_{ji} be the set of documents associated with category w_{ji} $(1 \leq i \leq k_j)$.
(3) Choose category w_{jh} with highest precision among W_j which has not yet been chosen. If $|D_{jh}| \leq Th$ then let $Result = Result \cup D_{jh}$. Otherwise, D_{jh} is further decomposed by each categorization viewpoint, let $D = D_{jh}$ and go to (2).
(4) Repeat (2) and (3) until (the recall of $Result$) $\geq r$. Output the precision and recall of $Result$.

Let $Rel(q)$ denote the set of relevant documents for a topic q. The precision and recall of $Result$ are defined as follows:

$$precision = \frac{|Result \cap Rel(q)|}{|Result|}, \quad recall = \frac{|Result \cap Rel(q)|}{|D_0 \cap Rel(q)|}.$$

To evaluate System B, entropy is used instead of clearness as the score in the above procedure. To evaluate System C, instead of step (2), K-means method is performed and let $D_{j1}, D_{j2}, \ldots, D_{jK}$ be the resultant clusters.

To evaluate System D, perform the following procedure for each topic q in the test set (Table 1). Let $Result = \emptyset$ and $Points = \emptyset$. For each topic q, specify a query which consists of one or two words chosen from the topic's name. Retrieve the ranked list $L = (d_1, d_2, \ldots, d_l)$ of documents by the query. For each $i = 1, 2, \ldots$, and l, let $Result = Result \cup \{d_i\}$ and add the precision-recall pair of $Result$ to $Points$. The precision $p(r)$ of System D for each recall r is defined as follows [16]:

$$p(r) = \max(\{p \mid (r', p) \in Points \text{ for } \exists r' \geq r\} \cup \{0\}).$$

Results and Discussion. The average precision-recall curves of ten topics are shown in Figure 5. In the figure, the curves labeled with Clearness, Entropy, Clustering and Keyword are those for Systems A, B, C and D, respectively. In Figure 5, we see that the results for Systems A and B are better than those for Systems C and D (on the average). This is because the proposed method uses *a priori* knowledge categorization viewpoints and can provide the decomposition which reflects the user's purpose of retrieval. When we examine the precision-recall curve for each topic in Table 1 independently, the same results hold for almost all topics. For low-recall, standard keyword approach (System D) performs almost as well as Systems A, B and C. However, the region where the recall is very low is not important since the recall is defined relative to not the entire document set but the document set retrieved by initial keywords for Systems A, B, and C.

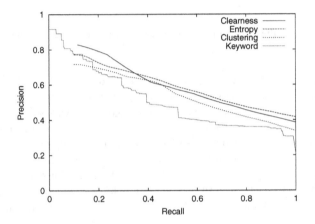

Fig. 5. Average Precision-Recall Curves

On the other hand, the results for System B (entropy) is slightly better than those for System A (clearness). The reason is as follows: if the number of documents associated with each category in a categorization viewpoint is smaller (the number of categories in the viewpoint is larger), then we can choose more relevant documents with the same recall and the precision becomes higher. By definition, the entropy of a categorization viewpoint S_j is higher when the number of documents associated with each category in S_j is smaller. As a result, a categorization viewpoint with higher entropy provides higher precision.

3.3   Experiment 2

In this experiment, the correlation of the categorization viewpoint score and the precision is evaluated for Systems A and B. The following is a detailed procedure for the experiment. W_j and D_{ji} are the same as those in Experiment 1.

For each topic q in the test set (Table 1) and for each $r = 0.2, 0.5$, and 0.8, perform the following (1) to (3):

(1) Retrieve the set D of documents by a few keywords for topic q. Calculate the score (clearness or entropy) of each categorization viewpoint for D.

(2) For each categorization viewpoint S_j, let $Result_j = \emptyset$ and repeat (a) until (the recall of $Result_j$) $\geq r$.

    a) Choose category w_{jh} with the highest precision among W_j which has not yet been chosen. Let $Result_j = Result_j \cup D_{jh}$.

(3) Arrange the categorization viewpoints $S_1, S_2, \ldots, S_m$ in the descending order of their score and in the descending order of their precision. Let the ordered lists be CV_s and CV_p, respectively. For each categorization viewpoint S_j $(1 \leq j \leq m)$, let $s(S_j)$ and $p(S_j)$ be the positions of S_j in the sorted lists CV_s and CV_p, respectively. Calculate the correlation coefficient of $s(S_j)$ and $p(S_j)$ with $1 \leq j \leq m$.

Results and Discussion. The results for Systems A and B are shown in Table 2. For System A, the precision and the score (clearness) have a positive correlation for 8 or 9 topics. For System B, the precision and the score (entropy) have a positive correlation for all topics. The reason why System B outperforms System A in correlation seems to be the same as for Experiment 1.

For each of the categorization viewpoints S_j with the top ten scores, the number of categories which are needed to satisfy (the recall of $Result_j$) $\geq r$ is examined. The average number of such categories among ten topics is also shown in Table 2. These average numbers for System A are less than those for System B. The two-way analysis of variance with independent variables, the score (clearness or entropy) and the topic, results in a significant difference between clearness and entropy in the case of $r = 0.5$ and $r = 0.8$ $(p < 0.05)$. Since it is desirable for a user to choose less categories to achieve the same precision, this result suggests that clearness is a better criterion of the score than entropy from a usability point of view.

Table 2. Averages of Correlation Coefficients between the Score and the Precision

Scoring criterion	r	The number of topics	$\overline{R}$ [*1]	The number of categories [*2]
Clearness	0.2	9	0.496	2.2
	0.5	8	0.595	4.1
	0.8	8	0.646	7.3
Entropy	0.2	10	0.596	2.3
	0.5	10	0.675	4.9
	0.8	10	0.684	8.7

[*1]: The average of the correlation coefficient between the score and the precision for each topic.
[*2]: The average number of categories which are needed to satisfy the recall $\geq r$ on each of the highest-scored ten categorization viewpoints.

4   Conclusion

In this paper, we proposed a method for supporting WWW retrieval using a flexible category structure. We also discussed experimental results made by a prototypic system based on the proposed method. The results showed that the proposed method provided higher precision than both the clustering method and the keyword-based method. It was also shown that scoring based on entropy provided slightly better precision than scoring based on clearness.

However, scoring based on clearness has the following advantages. First, the number of categories needed to achieve the same recall was smaller when scoring based on clearness than entropy. Second, the names of categorization viewpoints with higher clearness seemed to be more appropriate to the topic than the names with higher entropy (cf. Figure 6). This fact is important since a real user has to choose a categorization viewpoint and categories based on the category names without knowing the precision.

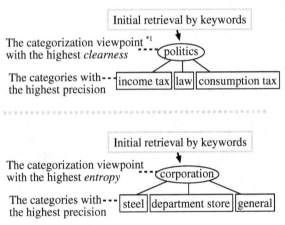

*1: Every categorization viewpoint and category is originally written in Japanese.

Fig. 6. Category Structures for the Topic 'Tax Reduction'

As future study, experiments using a human subject should be performed to improve the proposed method. For example, it should be clarified whether a user looses the advantage of being familiar with a fixed category when documents are dynamically divided into categories depending on initial keywords.

Acknowledgments. The authors sincerely thank Associate Professor Shin Ishii of Nara Institute of Science and Technology, for providing much instructive advice on the criteria for scoring the categorization viewpoints. The authors also thank Masahide Iwasaki for his assistance and Professor Dee A. Worman of Nara Institute of Science and Technology for carefully reading and editing the manuscript.

References

1. Anick, P. G. and Tipirneni, S.: The Paraphrase Search Assistant: Terminological Feedback for Iterative Information Seeking, in *SIGIR '99*, pp.153–159, 1999.
2. Dreilinger, D. and Howe, A. E.: Experiences with Selecting Search Engines using Metasearch, *ACM Trans. Information Systems*, Vol. 15, No. 3, pp.195–222, 1997.
3. Fishkin, K. and Stone, M. C.: Enhanced Dynamic Queries via Movable Filters, in *CHI '95*, pp.415–420, 1995.
4. Golovchinsky, G.: Queries? Links? Is there a difference?, in *CHI 97*, pp.407–414, 1997.
5. Grossman, D. A. and Frieder, O.: *Information Retrieval: Algorithms and Heuristics*, pp.134–142, Kluwer Academic Publishers, 1998.
6. Harada, M.: *Freya version 0.92*, 1998, http://odin.ingrid.org/freya/.
7. Kawano, H. and Hasegawa, T.: Data Mining Technology for WWW Resource Retrieval, in *IPSJ SIG Notes*, DBS108, pp.33–40, 1996.
8. Kitani, T., et al.: BMIR-J2 – A Test Collection for Evaluation of Japanese Information Retrieval Systems, in *IPSJ SIG Notes*, DBS114, pp.15–22, 1998.
9. Matsumoto, Y., Kitauchi, A., Yamashita, T., Hirano, Y., Imaichi, O. and Imamura, T.: Japanese Morphological Analysis System ChaSen Manual, Technical Report NAIST-IS-TR97007, Nara Institute of Science and Technology, 1997.
10. Pirolli, P., Shank, P., Hearst, M. and Diehl, C.: Scatter/Gather Browsing Communicates the Topic Structure of a Very Large Text Collection, in *CHI 96*, pp.213–220, 1996.
11. Pollitt, A. S.: The key role of classification and indexing in view-based searching, in *Proc. 63rd IFLA General Conf.*, 1997.
12. Robertson, G. G., Card, S. K. and Mackinlay, J. D.: Information Visualization using 3D Interactive Animation, *Comm. ACM*, Vol. 36, No. 4, pp.57–71, 1993.
13. Salton, G., Singhal, A., Buckley, C. and Mitra, M.: Automatic Text Decomposition Using Text Segments and Text Themes, in *Hypertext '96*, pp.53–65, 1996.
14. Sanderson, M. and Croft, B.: Deriving concept hierarchies from text, in *SIGIR '99*, pp.206–213, 1999.
15. Tou, J. T. and Gonzalez, R. C.: *Pattern Recognition Principles*, pp.89–97, Addison-Wesley, 1974.
16. Voorhees, E. M. and Harman, D. K.: Evaluation Techniques and Measures, in *The Seventh Text REtrieval Conference (TREC 7)*, p.A-1, National Institute of Standards and Technology (NIST), 1998.

Author Index

Lecture Notes in Computer Science

For information about Vols. 1–1844
please contact your bookseller or Springer-Verlag

Vol. 1882: D. Kotz, F. Mattern (Eds.), Agent Systems, Mobile Agents, and Applications. Proceedings, 2000. XII, 275 pages. 2000.

Vol. 1883: B. Triggs, A. Zisserman, R. Szeliski (Eds.), Vision Algorithms: Theory and Practice. Proceedings, 1999. X, 383 pages. 2000.

Vol. 1884: J. Štuller, J. Pokorný, B. Thalheim, Y. Masunaga (Eds.), Current Issues in Databases and Information Systems. Proceedings, 2000. XIII, 396 pages. 2000.

Vol. 1885: K. Havelund, J. Penix, W. Visser (Eds.), SPIN Model Checking and Software Verification. Proceedings, 2000. X, 343 pages. 2000.

Vol. 1886: R. Mizoguchi, J. Slaney /Eds.), PRICAI 2000: Topics in Artificial Intelligence. Proceedings, 2000. XX, 835 pages. 2000. (Subseries LNAI).

Vol. 1888: G. Sommer, Y.Y. Zeevi (Eds.), Algebraic Frames for the Perception-Action Cycle. Proceedings, 2000. X, 349 pages. 2000.

Vol. 1889: M. Anderson, P. Cheng, V. Haarslev (Eds.), Theory and Application of Diagrams. Proceedings, 2000. XII, 504 pages. 2000. (Subseries LNAI).

Vol. 1890: C Linnhoff-Popien, H.-G. Hegering (Eds.), Trends in Distributed Systems: Towards a Universal Service Market. Proceedings, 2000. XI, 341 pages. 2000.

Vol. 1891: Á.L. Oliveira (Ed.), Grammatical Inference: Algorithms and Applications. Proceedings, 2000. VIII, 313 pages. 2000. (Subseries LNAI).

Vol. 1892: P. Brusilovsky, O. Stock, C. Strapparava (Eds.), Adaptive Hypermedia and Adaptive Web-Based Systems. Proceedings, 2000. XIII, 422 pages. 2000.

Vol. 1893: M. Nielsen, B. Rovan (Eds.), Mathematical Foundations of Computer Science 2000. Proceedings, 2000. XIII, 710 pages. 2000.

Vol. 1894: R. Dechter (Ed.), Principles and Practice of Constraint Programming – CP 2000. Proceedings, 2000. XII, 556 pages. 2000.

Vol. 1895: F. Cuppens, Y. Deswarte, D. Gollmann, M. Waidner (Eds.), Computer Security – ESORICS 2000. Proceedings, 2000. X, 325 pages. 2000.

Vol. 1896: R. W. Hartenstein, H. Grünbacher (Eds.), Field-Programmable Logic and Applications. Proceedings, 2000. XVII, 856 pages. 2000.

Vol. 1897: J. Gutknecht, W. Weck (Eds.), Modular Programming Languages. Proceedings, 2000. XII, 299 pages. 2000.

Vol. 1898: E. Blanzieri, L. Portinale (Eds.), Advances in Case-Based Reasoning. Proceedings, 2000. XII, 530 pages. 2000. (Subseries LNAI).

Vol. 1899: H.-H. Nagel, F.J. Perales López (Eds.), Articulated Motion and Deformable Objects. Proceedings, 2000. X, 183 pages. 2000.

Vol. 1900: A. Bode, T. Ludwig, W. Karl, R. Wismüller (Eds.), Euro-Par 2000 Parallel Processing. Proceedings, 2000. XXXV, 1368 pages. 2000.

Vol. 1901: O. Etzion, P. Scheuermann (Eds.), Cooperative Information Systems. Proceedings, 2000. XI, 336 pages. 2000.

Vol. 1902: P. Sojka, I. Kopeček, K. Pala (Eds.), Text, Speech and Dialogue. Proceedings, 2000. XIII, 463 pages. 2000. (Subseries LNAI).

Vol. 1903: S. Reich, K.M. Anderson (Eds.), Open Hypermedia Systems and Structural Computing. Proceedings, 2000. VIII, 187 pages. 2000.

Vol. 1904: S.A. Cerri, D. Dochev (Eds.), Artificial Intelligence: Methodology, Systems, and Applications. Proceedings, 2000. XII, 366 pages. 2000. (Subseries LNAI).

Vol. 1906: A. Porto, G.-C. Roman (Eds.), Coordination Languages and Models. Proceedings, 2000. IX, 353 pages. 2000.

Vol. 1908: J. Dongarra, P. Kacsuk, N. Podhorszki (Eds.), Recent Advances in Parallel Virtual Machine and Message Passing Interface. Proceedings, 2000. XV, 364 pages. 2000.

Vol. 1910: D.A. Zighed, J. Komorowski, J. Żytkow (Eds.), Principles of Data Mining and Knowledge Discovery. Proceedings, 2000. XV, 701 pages. 2000. (Subseries LNAI).

Vol. 1912: Y. Gurevich, P.W. Kutter, M. Odersky, L. Thiele (Eds.), Abstract State Machines. Proceedings, 2000. X, 381 pages. 2000.

Vol. 1913: K. Jansen, S. Khuller (Eds.), Approximation Algorithms for Combinatorial Optimization. Proceedings, 2000. IX, 275 pages. 2000.

Vol. 1917: M. Schoenauer, K. Deb, G. Rudolph, X. Yao, E. Lutton, J.J. Merelo, H.-P. Schwefel (Eds.), Parallel Problem Solving from Nature – PPSN VI. Proceedings, 2000. XXI, 914 pages. 2000.

Vol. 1918: D. Soudris, P. Pirsch, E. Barke (Eds.), Integrated Circuit Design. Proceedings, 2000. XII, 338 pages. 2000.

Vol. 1920: A.H.F. Laender, S.W. Liddle, V.C. Storey (Eds.), Conceptual Modeling – ER 2000. Proceedings, 2000. XV, 588 pages. 2000.

Vol. 1921: S.W. Liddle, H.C. Mayr, B. Thalheim (Eds.), Conceptual Modeling for E-Business and the Web. Proceedings, 2000. X, 179 pages. 2000.

Vol. 1922: J. Crowcroft, J. Roberts, M.I. Smirnov (Eds.), Quality of Future Internet Services. Proceedings, 2000. XI, 368 pages. 2000.

Vol. 1923: J. Borbinha, T. Baker (Eds.), Research and Advanced Technology for Digital Libraries. Proceedings, 2000. XVII, 513 pages. 2000.

Vol. 1924: W. Taha (Ed.), Semantics, Applications, and Implementation of Program Generation. Proceedings, 2000. VIII, 231 pages. 2000.

Vol. 1926: M. Joseph (Ed.), Formal Techniques in Real-Time and Fault-Tolerant Systems. Proceedings, 2000. X, 305 pages. 2000.

Vol. 1927: P. Thomas, H.W. Gellersen, (Eds.), Handheld and Ubiquitous Computing. Proceedings, 2000. X, 249 pages. 2000.

Vol. 1931: E. Horlait (Ed.), Mobile Agents for Telecommunication Applications. Proceedings, 2000. IX, 271 pages. 2000.

Vol. 1766: M. Jazayeri, R.G.K. Loos, D.R. Musser (Eds.), Generic Programming. Proceedings, 1998. X, 269 pages. 2000.

Vol. 1933: R.W. Brause, E. Hanisch (Eds.), Medical Data Analysis. Proceedings, 2000. XI, 316 pages. 2000.